GENETIC MEDICINE AND ENGINEERING

Ethical and Social Dimensions

Edited by

Albert S. Moraczewski, OP, PhD
Vice President for Research
The Pope John XXIII Medical-Moral Research and
Education Center

The Catholic Health Association of the United States
St. Louis, MO

and

The Pope John XXIII Medical-Moral Research and Education Center
St. Louis, MO

Nihil Obstat
　　Robert F. Coerver

Imprimatur:
　　E. J. O'Donnell
　　Vicar General of St. Louis
　　March 8, 1983

Copyright 1983
by
The Catholic Health Association of the United States
St. Louis, MO 63134
and
The Pope John XXIII Medical-Moral Research and Education Center
St. Louis, MO 63134

Printed in the United States of America. All rights reserved.
This book, or any part thereof, may not be reproduced
without the written permission of the publisher.

Library of Congress Cataloging in Publication Data
Main entry under title:

Genetic medicine and engineering.

　Papers presented at three seminars entitled, Genetics and health care, held in the fall of 1981.
　Includes bibliographical references and index.
　1. Medical genetics—Moral and ethical aspects—Congresses.　2. Genetic engineering—Moral and ethical aspects—Congresses.　3. Medical genetics—Law and legislation—Congresses.　I. Moraczewski, Albert S., 1920-　　　II. Catholic Health Association of the United States. III. Pope John XXIII Medical-Moral Research and Education Center. [DNLM:　1. Ethics, Medical—Congresses.　2. Genetics, Medical—Congresses.　3. Genetic intervention—Congresses.　QZ 50E83 1981]

RB155.G386　1983　　174'.2　　　82-22019
ISBN 0-87125-077-2

Contents

Authors / v
Foreword / vii
Preface / xi
Acknowledgements / xv
Part I: Genetic Medicine / 1
    1 Genetic Disorders / 3
       Kutay Taysi, MD
    2 Genetic Diagnosis / 15
       Rev. Albert S. Moraczewski, OP, PhD
    3 Genetic Counseling / 31
       Rev. Robert Baumiller, SJ, PhD
    4 Genetic Screening and Prenatal Diagnosis / 45
       Robert F. Murray, Jr, MD
Part II: Genetic Engineering / 61
    5 Treatment of Genetic Diseases: Inborn Errors of Metabolism / 63
       Seymour Packman, MD
    6 An Introduction to Genetic Engineering / 75
       H. Michael Shepard, PhD
Part III: Ethical and Social Aspects / 85
    7 Ethical Principles and Genetic Medicine / 87
       Rev. Donald G. McCarthy, PhD
    8 Genetic Manipulation: Some Ethical and Theological Aspects / 101
       Rev. Albert S. Moraczewski, OP, PhD
    9 Legal Issues Relating to Genetic Diagnostic Procedures / 121
       G. Edward Fitzgerald, LLB
   10 Genetic Disease, Counselors, and the Wider Society: A Philosopher's View / 131
       Gary M. Atkinson, PhD
   11 Social Implications of Genetic Manipulation / 145
       Rev. Daniel J. Sullivan, SJ, PhD
Epilogue / 161
Glossary / 165
Bibliography / 179
Index / 187

Authors

Gary M. Atkinson, PhD, Department of Philosophy, College of St. Thomas, St. Paul, MN.

Rev. Robert Baumiller, SJ, PhD, director, Division of Genetics, Department of OB/GYN, Georgetown University Hospital, Washington, DC.

Sr. Betty Brucker, SSM, executive director, St. Mary's Health Center, St. Louis, MO.

G. Edward Fitzgerald, LLB, Gibson, Dunn and Crutcher, Los Angeles, CA.

Rev. Donald G. McCarthy, PhD, St. Cecilia's Parish, Cincinnati, OH, and director of education, The Pope John XXIII Medical-Moral Research and Education Center, St. Louis, MO.

Rev. Albert S. Moraczewski, OP, PhD, vice president for research, The Pope John XXIII Medical-Moral Research and Education Center, St. Louis, MO.

Robert F. Murray, Jr, MD, chief, Division of Medical Genetics, Department of Pediatrics and Child Health, Howard University College of Medicine, Washington, DC.

Seymour Packman, MD, Department of Pediatrics, Division of Genetics, School of Medicine, University of California, San Francisco, CA.

H. Michael Shepard, PhD, Department of Molecular Biology, Genentech, Inc., South San Francisco, CA.

William S. Sly, MD, professor of Pediatrics, Genetics, and Medicine, Washington University School of Medicine, and director, division of Medical Genetics, St. Louis Children's Hospital, St. Louis, MO.

Rev. Daniel J. Sullivan, SJ, PhD, associate professor, Department of Biological Sciences, Fordham University, Bronx, NY.

Kutay Taysi, MD, director, Medical Genetics Clinic and Cytogenetics Laboratory, St. Louis Children's Hospital, St. Louis, MO.

Foreword

Genetics has clearly moved to center stage in biology, in biochemistry, and in medicine. The dramatic implications for medical and hospital practice unfold weekly, if not daily, in lay newspapers and magazines as well as medical journals. I think it is important, in setting the stage for the subject matter discussed in this book, to review briefly how this all came to pass.

Barton Childs recently reviewed the development of educational efforts in genetics in medical schools in the United States. In 1945, only 7 of 84 American medical schools had courses in genetics. In 1955, the number was still only 7 of 87 schools. By 1975, however, 84 of 103 schools listed formal genetics courses in their curricula. Three years later, this number had increased to 103 of 107 schools polled. It is very clear from these numbers that, between 1955 and the present, medical genetics came almost out of nowhere to be perceived universally as an essential part of a physician's training. Several factors contributed to this dramatic development of genetic medicine.

One important factor was completely outside genetics. This was the dramatic progress in the control of environmentally induced diseases over the first 60 years of this century. As smallpox, rheumatic fever, tuberculosis, and infantile diarrhea lost their status as big killers in developed countries, genetic abnormalities and birth defects, which had undergone no absolute increase in prevalence, steadily increased in relative importance as causes for hospitalization and death. By 1960, it seemed clear, to some at least, that training specialists to handle this increasingly important component of human disease was worthwhile, and the first postgraduate training programs in medical genetics were begun. This effort focused more attention on inherited diseases, which became increasingly widely recognized and better characterized. The steady expansion of recognized genetic disorders, from 412 in 1958 to 3,368 in the latest (1982) *McKusick Catalogue of Mendelian Inheritance,* demonstrates the impact of increasing recognition of inherited diseases in human beings. This understates the importance of genetic factors in medicine, however, since this impressively large catalogue does not include the growing list of genetic factors that contribute significantly, in still unexplained ways, to very common disorders such as coronary heart disease, diabetes mellitus, schizophrenia, hypertension, and allergic disorders. This inherited susceptibility to a variety of disorders is just now beginning to attract widespread attention.

The 1960s also saw chromosome studies move into the realm of medical practice. Nearly 1 of every 200 newborns has a detectable chromosome abnormality. The clinical consequences of these chromosome imbalances have made chromosome studies important diagnostic tools in problems involving birth defects, mental retardation, multiple miscarriages, and infertility. Over the last 2 decades the techniques have become increasingly sensitive. Physicians and scientists are very near perfecting the tools to combine recombinant deoxyribonucleic acid (DNA) techniques with chromosome studies to locate single-gene defects in chromosome preparations.

Another dramatic development was the application of the techniques of amniocentesis, ultrasound, and, more recently, fetoscopy to the diagnosis of genetic disorders in the second trimester of pregnancy. Amniocentesis in particular proved far safer than expected and has become a nearly standard method for prenatal diagnosis of chromosomal and biochemical genetic disorders. The availability of selective abortion of infants found to have genetic abnormalities has been widely touted as a means of "preventing genetic diseases." Naturally, this is not an acceptable alternative for many. More often, however, the prenatal diagnostic studies provide information that reassures parents and allows them to continue their pregnancy without dread of the outcome related to their genetic risk.

These are some of the developments that led genetics to demand attention from medical educators and physicians. Provision for delivery of genetic services (diagnostic services and counseling for genetic risks) has come to be expected in any tertiary medical center. Neither primary physicians nor community hospitals, however, can ignore their responsibility to be aware of which patients might benefit from these services and to advise patients who need these services of their options. Several recent court decisions have cleared the way for medical malpractice suits against primary physicians for failure to provide genetic counseling to patients with higher than random risks for genetic disease.

Since 1975 a whole new dimension has been added. The advent of gene splicing and recombinant DNA techniques generated a whole new industry called genetic engineering. Growth hormone, produced in bacteria after transfer of cloned human growth hormone genes to bacteria, provides a striking example of how harnessing these tools to solve specific medical problems holds great promise. The world shortage of growth hormone for patients with inherited or acquired deficiency of this hormone has been solved through this breakthrough. Equally dramatic instances of use of these tools to explore the structure and function of genes, to define the genes responsible for many forms of human cancer, and to define the mechanisms for many kinds of inherited diseases are either already in the headlines or about to make news.

It is a wonderful time to be a geneticist. Each new development generates new opportunities. Clearly, some developments generate important practical problems and raise new ethical issues as well. This book seeks to explore those issues. Although this book tries to capture the importance of the genetic revolution to medicine, it should not be considered the last word. The revolution is not over; it is still actively developing. This book also has a unique dimension. It is the only book I know of that tries to view these developments from the vantage point of those with responsibilities as decision makers in the health care profession.

William S. Sly, MD
Professor of Pediatrics, Genetics, and Medicine
Washington University School of Medicine
Director, Division of Medical Genetics
St. Louis Children's Hospital

Preface

What do genes, genetic engineering, and the book of *Genesis* have in common? Genes are the units of inherited characteristics. Genetic engineering is the developing technology by which scientists increasingly are able to manipulate genes and thus design new forms of life and radically correct inherited disorders of plants, animals, and humans. The book of *Genesis* provides a religious description of human origins and basic guidance on the control human beings exercise over their inherited characteristics.

With the discovery of the Augustinian monk Gregor Mendel in 1866 that plants, and presumably all living things, transmitted certain characteristics from one generation to another by discreet packets or units, the field of genetics was born, even if inauspiciously, since no one at that time recognized the significance of Mendel's paper. The rediscovery of this critical bit of biological knowledge, by three botanists in 1900, led more quickly to public recognition of this phenomenon.

During the subsequent decades new information about genes was gradually added, but not until the 1950s was there sufficient evidence to conclude that the biochemical correlate of genes was the biochemical substance deoxyribonucleic acid (DNA). In 1953 Watson and Crick discovered that DNA occurred in long strands shaped in the form of a double helix. This knowledge provided a basis for understanding the manner in which the chromosome—the bearer of genes, composed largely of DNA plus a protein—duplicated itself. At the same time the process by which DNA influenced or directed the synthesis of proteins became better understood. An essential step in learning the exact biochemical steps involved in transmitting the information stored in DNA molecules—each containing perhaps thousands of genes in the human being—was the discovery of the "genetic code." The DNA molecule is made up of four chemical compounds, called nucleotides, which encode various bits of information by the sequence in which they are connected, generally in groups of three. These groups constitute the information unit or letter of the gene, somewhat like the dots and dashes of Morse code (a two-symbol code) stand for letters.

In addition, two significant developments occurred concomitantly. One was scientists' increasing ability to construct and modify the DNA molecules of living organisms, particularly of such simple forms of life as bacteria, and to insert such altered genes into an organism. Such genetic modification gave the modified organism characteristics it did not naturally possess; it could, for example, synthesize substances totally foreign to it, such as *Escherichia coli* acquiring by genetic alteration the ability to produce human insulin or growth hormone.

The other development was an increasing understanding of the genetic contribution to numerous diseases and disorders. The fifth edition of McKusick's catalogue lists over 3,000 disorders that have a genetic component. Concurrently, the various modes of inheritance were more fully appreciated. Such knowledge permitted the medical geneticist to make quantitative predictions on whether a particular disorder would occur or reoccur in a given family.

As the various basic science disciplines and clinical sciences relevant to genetics developed, a recognized medical speciality emerged. In fall 1981 the first round of National Board Examinations was given in the United States to certify genetic counselors. Such certification will reduce the opportunity for unqualified persons to enter this very complex and sensitive field.

The burgeoning of genetic knowledge, the increased capabilities of genetic therapy, and the greater number of qualified genetic counselors has now made it necessary for health care facilities in general, and Catholic health care facilities in particular, to consider seriously including genetic counseling, diagnosis, and testing in their repertory of services. The relative newness and complexity of the field poses some new legal and some thorny ethical problems for boards of trustees and administrators, however.

In light of these considerations, a series of seminars, "Genetics and Health Care," was sponsored by the Knights of Malta and The Catholic Health Association of the United States (CHA) in association with the Pope John XXIII Medical-Moral Research and Education Center which developed the program. These invitational seminars on genetics, delivered in fall 1981, were addressed to the major decision makers of the Church and of the Catholic health care ministry in the St. Louis region. The participants found them very helpful and suggested that the material be made available to a larger audience. Reinforced by such affirmation, CHA and the Pope John Center decided to publish jointly the lectures and discussions.

Accordingly, this book intends—as did the seminars on which it is based—to assist the major decision makers in the Catholic health care ministry (e.g., congregational leaders, facility administrators, diocesan personnel) in formulating appropriate and practical policies in matters of genetic medicine. For example, should a Catholic health care facility maintain a genetic counseling clinic? Can it provide prenatal diagnostic procedures such as amniocentesis? Does a hospital become legally vulnerable if it refuses to offer abortion advice in a genetic counseling clinic? Are there specific guidelines for trying experimental procedures in gene therapy? How realistic are the possibilities of therapy for genetic disorders? From the viewpoint of the Church's teaching, may scientists alter a person's genetic make-up? All these and other concerns are considered in this book.

In the past few years a number of works on various aspects of genetic medicine have appeared. Some are clearly directed to the expert: the geneticist or genetic counselor. Others are intended for the general lay public. But few have in mind the hospital administrator or trustee, i.e., those persons who are in responsible decision-making positions but do not have an expert's knowledge of medical genetics, although they have varying degrees of medical and scientific knowledge. This work is intended to assist these health care professionals in their decision-making responsibilities.

At the same time, the needs of physicians, nurses, and various specialists in medical genetics have not been overlooked. The interdisciplinary nature of this book will aid the individual professional in appreciating more fully some of the wider dimensions of genetic medicine and engineering.

The reader should note that this book may be considered a companion volume to a 1980 publication of Pope John Center, *Genetic Counseling, The Church and The Law*. That work was the report of a Pope John Center's research task force on genetics in medicine, and to some extent this book balances that work's scope. This book does not review fundamental genetics, consider *in extenso* the Biblical and theological meaning of the human being, or address in detail the subject of tort law and genetics, all of which are covered by the first book. It does discuss a number of genetic diseases not mentioned in the earlier volume, however, and it deals in more detail with some diagnostic techniques. Furthermore, it grapples with recombinant DNA technology and sketches some theological reflections on genetic engineering. One chapter is devoted to the *treatment* of genetic diseases, focusing particularly on inborn errors of metabolism. The legal discussions are updated, and one chapter reflects the thoughts of an attorney who was involved with a famous California case, *Curlender v. Bio-Science Laboratories*. Both volumes seek to address the multifaceted issues against the background of Christian values and the Church's teachings. To enhance the value of these two works, this book has a combined index which covers the material in both volumes.

Hence, the book aims to assist the reader as follows:
- To appreciate the current status of scientific research and medical practice in genetics;
- To understand the legal and ethical implications of current practice and research in genetics;
- To relate current genetic knowledge and technology to the roles and responsibilities of decision makers in Catholic health care facilities;
- To identify basic guidelines for decisions relating to genetic diagnoses, counseling, and treatment in Catholic facilities; and
- To consider the implications of future developments in genetic practice and research.

The chapters are arranged into three groups. Part I examines genetic diagnosis, counseling, and screening. Part II considers genetic engineering in a broad sense so as to include manipulation of genes or of the product of their activities for purposes of treatment. Legal, ethical, and social aspects of genetic medicine and engineering are discussed in Part III. An epilogue reflects on some additional implications for Catholic health care facilities.

> Rev. Albert S. Moraczewski, OP, PhD
> Editor

Acknowledgements

Besides the lecturers and the participants in the discussions, many other persons and organizations must be gratefully acknowledged. First, I wish to extend sincere thanks to the American Association of the Knights of Malta, whose generous grant made the seminars upon which this book is based possible. Special gratitude is owed to Daniel L. Schlafly, former regional vice president of the American Association of the Knights of Malta, whose personal encouragement was ever present. Deeply felt appreciation is also extended to the Most Rev. Theodore E. McCarrick, former chaplain of the American Association of the Knights of Malta, and to the Most Rev. John L. May, archbishop of St. Louis.

During the development of the seminars, much help was given by Patricia Monteleone, MD, professor of pediatrics, St. Louis University School of Medicine, and director, Division of Medical Genetics, Cardinal Glennon Memorial Hospital for Children, and by William Sly, MD, professor of pediatrics, medicine and genetics, Washington University School of Medicine, and director, Division of Medical Genetics, St. Louis Children's Hospital. Both, in addition, read a number of the papers and made valuable suggestions.

Further excellent advice was received from the Rev. Benedict Ashley, OP, PhD, STM, professor of moral theology, Aquinas Institute, who read a substantial part of the book in draft form.

Any book, even if slim and modest in its endeavors, nonetheless is at times a great burden for editor (or author). Accordingly, thanks must also be expressed to those whose personal and official support not only contributed to the seminars but permitted the book to see the light of print. Among these outstanding supporters are John E. Curley, Jr., president, CHA, and the Rev. William M. Gallagher, president, the Pope John Center. Center stage, however, must be given to CHA's Sr. Margaret John Kelly, DC, PhD, vice president, Mission Services, who shared the responsibility of conducting the seminars and was most helpful and supportive in completing the task of publishing the book.

Finally, proper tribute must be paid to Mrs. Nancy Hoffman and Mrs. Lucie Smith, who carried out with distinction the laborious task of transcribing the lectures and discussions from the audiotapes. Mrs. Lucy Diefenbach patiently typed and retyped various drafts of the manuscript. Members of CHA's publications department also offered valuable assistance.

To all the above persons, and any whose name I may have inadvertently omitted, I express my grateful thanks in the Lord.

Rev. Albert S. Moraczewski, OP, PhD
Editor

PART I

GENETIC MEDICINE

chapter 1

Genetic Disorders

Kutay Taysi, MD

Genetic disorder, or genetic disease, can be defined as a disease that is caused by abnormal genetic material. Abnormal genetic material may be a single abnormal gene, multiple abnormal genes, or abnormal chromosomes (either an abnormal number or an abnormal structure). There are thus three different types of genetic diseases. This chapter will present an overview of these three types of diseases and, in so doing, provide a foundation for the subsequent chapters.

Medical genetics, which deals with hereditary disorders, is a relatively new discipline. Genetic disorders became the focus of attention only recently. What is the reason for this recent interest? Is there an epidemic of genetic disorders? Certainly not. The answer is very simple. No evidence indicates an increase in the frequency of genetic disorders, but the relative frequency and the relative importance of genetic disorders have increased in the last few decades.

At the beginning of the 20th century, U.S. infant mortality was 150:1,000 live births. Approximately 5 of these 150 deaths were thought to be related to genetic causes. In other words, almost 80 years ago, only 3.3 percent of all infant deaths was thought to be due to genetic causes. Since then, dramatic improvement in infant nutrition and social and environmental factors, and, most important, the introduction of antibiotic therapy have decreased infant mortality considerably. Today U.S. infant mortality is 15:1,000—an almost 90 percent reduction. Of these 15 infant deaths for each 1,000 live births today, however, 5 still are related to or are the result of genetic or partly

---

Dr. Taysi is director, Medical Genetics Clinic and Cytogenetics Laboratory, St. Louis Children's Hospital, St. Louis, MO

genetic causes.[1] Thus, today 33 percent of infant mortality is related to genetic causes, whereas at the beginning of the century, only 3.3 percent of the infant mortality was related to genetic causes. This relative increase is found not only in infant mortality but in infant morbidity as well.

Many people, including many physicians, believe that genetic diseases are very rare. That is a misconception. Recent data from the United States and Canada show that approximately one third of all pediatric admissions to childrens' hospitals are for birth defects, i.e., for genetic diseases or diseases with some degree of a genetic component in their etiology.[2] Furthermore, patients with genetic diseases enter hospitals and stay longer, at least 5 times as frequently as patients with nongenetic diseases.[3,4] Therefore, it is apparent that although individual genetic disorders may be relatively uncommon, the sum total has a significant effect on mortality and the morbidity at all ages, especially in infants and children.

Basic Genetics

As mentioned above, genetic diseases are caused by abnormal genetic material: a single abnormal gene (a mutant gene), multiple abnormal genes, or abnormal chromosome(s). A gene is a segment of the deoxyribonucleic acid (DNA) molecule, which determines a specific inherited characteristic. Genes are organized in structures called chromosomes, which are located in the nucleus of most cells (one exception is, for example, mature red blood cells). In a biochemical sense, genes determine the production of proteins, such as enzymes. If a particuiar gene is abnormal, or mutant, enzyme production could be deficient or absent, which most often will cause a clinical problem.

All the genetic material, or genetic information, inherited from both parents is present in the first cell of the fetus. The first cell, the fertilized egg, also called a zygote, has the full complement of 46 chromosomes. It is a multipotent cell, which means it has the capacity to produce any of the more specialized cells subsequently needed. From this single cell, in approximately 9 months, a human being will be born. How does it happen?

During the development of the embryo and fetus, two important events take place: cell division and cell differentiation. The newborn infant has an enormous number of cells (2×10^{14}) as a result of cell *division* during development. Furthermore, a newborn infant has many cells with different morphology that serve different functions. The original cell (the fertilized egg) and all the cells derived from it have exactly the identical and full complement of genetic material. Then why does a newborn infant have different cells, e.g., kidney, liver, and skin cells? Different cells are due to cell *differ-*

entiation. In all cells, the total genetic material, or the genome, is the same, except for the reproductive cells (egg and sperm), each of which have only half the number of chromosomes. During embryonic or fetal development, however, certain genes in particular cells function and other genes do not; certain genes are turned on and others are turned off. This is the fundamental mechanism of cell differentiation, by which billions of cells in the human body with diverse functions and diverse morphology are developed.

The development of the embryo and fetus is precisely programmed in a chronological and sequential manner. Certain developmental stages should be reached at certain times. Any disruption of this carefully timed sequential and extremely delicate process will result in birth defects. Such disruption can be caused by several factors: some purely genetic, some purely environmental, and some a combination. Since this chapter focuses on genetic diseases, I will only mention briefly those birth defects caused by nongenetic agents harmful to fetus, i.e., teratogens.

Causes of Congenital Malformations (Birth Defects)

Causes of birth defects can be divided into four groups:
- Single-gene diseases (diseases with mendelian inheritance)
- Chromosome disorders
- Multifactorial diseases (polygenic diseases)
- Diseases caused by teratogens

The first two groups include purely genetic diseases. The third group of diseases is caused by the combination of genetic and nongenetic, environmental factors. Finally, the fourth group is caused by environmental factors.

Single-Gene Disorders

There are approximately 3,000 different diseases in the single-gene-disorder group, all of which are caused by different abnormal (mutant) genes.[5] A gene mutation is a chemical change in gene structure. The reason for gene mutation is largely unknown. Once a gene becomes mutant, however, it is transmitted in its mutant form from one generation to the next. Therefore, single-gene diseases run in families and follow certain predictable types of inheritance.

Smith-Lemli-Opitz Syndrome. One example of a single-gene disorder is Smith-Lemli-Opitz syndrome.[6] This disorder is due to a single abnormal autosomal (i.e., not carried on the sex chromosomes) recessive gene. The

---
*A genetic disorder that requires that both parents (nonaffected and carrier) transmit the defective gene to the offspring.

abnormal gene in some way disrupts the developmental program of the fetus and results in a well-defined clinical entity. Patients with Smith-Lemli-Opitz syndrome have severe growth and mental retardation, feeding problems in infancy, and typical facial features, all of which permit the physician to make the appropriate clinical diagnosis. Physicians do not know the chromosomal location of this abnormal (mutant) gene, nor do they know the basic abnormality that the mutant gene causes at the molecular level. The clinical disorder may be caused by an enzymatic deficiency, because an enzyme deficiency usually exists in most autosomal recessive* conditions. Although little is known about the primary effect of this abnormal gene, parents still receive counseling because the inheritance pattern is known to be autosomal recessive. Parents with a child who has Smith-Lemli-Opitz syndrome have a 25 percent chance of having another child with the same condition.

Meckel-Gruber Syndrome. Another example of a single-gene disorder is Meckel-Gruber syndrome.[7] The main features of this severe condition are encephalocele (herniation of the brain tissue and central nervous system fluid in a bulging sac through a hole in the skull), cysts in the kidneys (polycystic kidneys), and extra fingers. Because of the severe brain and kidney problems, these infants do not survive more than a few days or weeks. The correct diagnosis is extremely important for genetic counseling, however. The following examples will clarify this point. Very similar defects can also be found in a chromosomal condition called trisomy 13 (an extra chromosome no. 13 in each cell).[8] The prognosis of trisomy 13 is very poor. Clinical features in both conditions are quite similar, but the recurrence rate of trisomy 13 is not generally more than 1 percent. An encephalocele can be found as an isolated malformation in an infant, with a recurrence risk approximately between 3 percent and 5 percent in the same family. By comparison, the recurrence risk in Meckel-Gruber syndrome is 25 percent. Meckel-Gruber syndrome can be diagnosed prenatally through special prenatal diagnostic tests. Parents can thus be counseled on their high risk and can receive prenatal diagnosis for this very serious condition.

Lesch-Nyhan Syndrome. Another example of a single-gene disorder that causes birth defects is Lesch-Nyhan syndrome.[9] This disorder is much better understood, since physicians know the basic biochemical defect responsible for the clinical problems. In Lesch-Nyhan syndrome, an abnormal gene is located on the X chromosome (called X-linked recessive gene). The disorder is responsible for the deficient production of the enzyme Hypoxanthine

---

*A genetic disorder that requires that both parents (nonaffected and carrier) transmit the defective gene to the offspring.

Guanine Phosphoribosyl Transferase (HGPRT), which results in an unregulated purine synthesis. Patients with Lesch-Nyhan syndrome are healthy at birth; however, at around 8 to 10 months of age, they start to have neurological problems such as spastic paralysis of the lower legs, abnormal arm and leg movements (choreoathetosis), mental retardation, and self-mutilation (they bite their lips and fingers). The whole Lesch-Nyhan story is a very remarkable example of the rapid development of medical understanding of a genetic disorder. In less than a 10-year period, physicians were able to locate the gene on the X chromosome, to understand the biochemical defect, and to develop tests for carrier detection and prenatal diagnosis. Now carriers can be counseled on their increased risk, which is 25 percent (i.e., half the sons of a carrier mother will be affected), and pregnant carriers can take the prenatal diagnostic test.

Achondroplasia. Another example of a single-gene disorder as a cause of birth defects is achondroplasia.[10] In this disorder, a single dominant* gene disrupts the precise developmental program of the fetus. Dominant genes differ from recessive genes in that the dominant mutation generally affects structural proteins, causing structural problems, whereas recessive mutation usually causes enzymatic deficiency. The main problem in these patients is their short stature because of skeletal abnormality. Intelligence is normal. Affected persons have a 50 percent risk of having similarly affected children. Achondroplasia can be diagnosed by a physical examination and by x-ray studies in the newborn period rather easily.

Chromosomal Disorders

The main features of the chromosomal disorders are growth and mental retardation, developmental brain defects, and congenital heart diseases. Affected persons also have unusual facial features due to abnormal skull development.

In single-gene disorders, the defect is a *qualitative* defect in the gene function. The main problem is a chemically abnormal gene that is causing abnormal gene function. In chromosomal disorders, the quality of the genes is normal. Chromosome abnormality is a *quantitative* problem. Either one chromosome too many or one too few is present in each body cell. This quantitative abnormality in the gene function in some way disrupts the developmental program of the fetus and may result in birth defects and mental retardation.

Chromosomal abnormalities are a frequent cause of miscarriage, neonatal

---

*Dominant gene conditions usually require that only one of the parents transmit a single mutant gene to the offspring.

morbidity, death in the first year of life, and mental retardation. Sixty percent of all fetuses that are miscarried in early pregnancy, and 6 percent of stillbirths, have chromosomal abnormalities. The frequency of chromosomal abnormality in the liveborn infants is 0.6 percent.

The most common chromosomal problem in the population is Down syndrome, or trisomy 21.[11] Affected patients have typical clinical features. They have one extra chromosome no. 21 in each of their body cells, so that the total number of chromosomes is 47 instead of the normal 46. This extra chromosome, in a way not yet understood, disrupts the developmental program of the fetus, causing serious birth defects and mental retardation.

Most chromosomal disorders are not hereditary. The recurrence risk is 1 percent or less in families with a single affected child. Some chromosomal problems may be hereditary, however, with a recurrence risk between 10 and 100 percent. Certainly, a 100 percent risk is extremely rare, but it may happen. In order to determine which chromosomal problem is hereditary and which is not, a chromosome analysis of the affected person and, if necessary, of the parents is required.

Multifactorial (Polygenic) Diseases

Most of the common birth defects, such as congenital heart defects, cleft lip and palate, and open spine defects (neural tube defects), appear to run in families.[12] These disorders are not associated with chromosomal problems and are not due to a single abnormal gene. In this group of disorders, the precise nature of the genetic component is not clear. Apparently, both genetic and environmental nongenetic factors play a role in etiology. These disorders are thus said to have multifactorial inheritance because both genetic and nongenetic factors are involved. Apparently the genetic component is more than one abnormal gene. The term *polygenic inheritance* is therefore also used for this inheritance pattern. In families with one affected member, the recurrence risk is between 2 percent and 5 percent. Furthermore, prenatal diagnosis is available for open spine defects (e.g., myelomeningocele).

Diseases Caused by Teratogens

The fourth group of birth defects is not one of genetic disorders. These are the diseases caused by teratogens—substances or factors inducing birth defects. These agents harm the developing fetus. A brief summary of well-known teratogenic agents and their effect on developing fetuses follows. Although these defects do not have a genetic cause, it would be incomplete not to mention, albeit very briefly, some common teratogens and their effects.

One of the most common examples is maternal rubella infection (German measles).[13] Some pregnant women exposed to the rubella virus during early pregnancy will have infants with microcephaly (abnormally small head), mental retardation, congenital heart disease, and deafness.

The second most important, and most widely known, teratogen is x-ray radiation.[14] Although diagnostic x-ray examination in small doses during pregnancy is not harmful, large doses, especially therapeutic radiation during pregnancy, will cause birth defects and mental retardation. The most typical effect of maternal heavy radiation is microcephaly, which in most cases causes mental retardation.

Many chemical substances have been implicated in causing birth defects when used by pregnant women. The well-known antiepileptic drug, diphenylhydantoin (Dilantin), for example, can cause significant birth defects in the exposed fetus.[15] The main features of babies exposed to diphenylhydantoin in utero are growth and mental retardation, cleft lip and palate, short nose, abnormal ears, and hypoplastic (underdeveloped) fingernails and toenails.

Chronic heavy alcohol intake is another big problem.[16] It is very harmful to the fetus. A portion of the infants whose mothers are chronic, heavy drinkers will show typical facial features in addition to growth and mental retardation. Most babies with full blown clinical features of the fetal alcohol syndrome have been born to women whose alcohol intake is 8 to 10 hard drinks per day. Physicians know little about the effect of small amounts of alcohol, but recent evidence suggests that it too may be a problem, even if not as severe a one as the full-blown syndrome of heavily and chronically addicted mothers (fetal alcohol syndrome).

Recent Developments

After reviewing the types of genetic disorders and factors causing congenital malformations, I would like to summarize the most important developments that helped bring medical genetics to the daily practice of medicine.

The first factor in this development is the increased relative frequency and relative importance of the genetic disorders in the population as a result of a reduction of nongenetic diseases.

The second factor is the improved cytogenetic technology, such as the new chromosome methods, which increases physicians' ability to detect small chromosomal abnormalities that had been missed in the past.

A third factor is the development of powerful prenatal diagnostic methods such as ultrasonography and amniocentesis as a part of genetic counseling.

(Parenthetically, I wish to make one brief comment about prenatal diagnosis, since it will be discussed in much more detail in the following chapter. Virtually all chromosome abnormalities in the fetus, more than 100 biochemical defects, neural tube defects [open spine defects], a number of other congenital malformations, and certain hemoglobin disorders can now be diagnosed by prenatal diagnostic methods.)

The fourth significant factor in the development of medical genetics is the advent of screening for genetic disorders. This subject will be treated at greater length in Chapter 4. Before the advent of genetic screening, physicians would counsel parents for their recurrence risk *after* they had had an affected child. Physicians would then inform them of their high risk of having another affected child. Now that methods exist to screen couples at high risk for Tay-Sachs disease, for example, if physicians determine that both parents are carriers, they can inform the parents of the 25 percent risk. This is *prospective* counseling, rather than *retrospective* counseling.

With an understanding of the basic genetic concepts and a knowledge of recent technology, medical professionals and health care administrators can improve the delivery and the quality of genetic health care services. In this way, the tragic emotional and financial impact of serious genetic diseases on a family and on society can be minimized.

Discussion

Q Dr. Taysi, you said that genetic disorders cause a rather substantial amount of inpatient days at pediatric hospitals. Because you did not say the same about "adult" hospitals, it seems that most of these problems were corrected. Is it true that most genetic disorders that result in inpatient stays at pediatric hospitals are corrected?

A In a sense that is correct. I said "corrected" mainly because of the high mortality in chromosomal disorders. Most of these patients die, unfortunately, before they reach adulthood. With improved diagnostic and medical care, physicians are starting to see almost all the genetic abnormalities, even chromosomal disorders, in all age groups. There are adult patients with chromosomal disorders, but they are rare. On the other hand the number of adult patients with multifactorially inherited group, such as heart disease, diabetes, and schizophrenia is high.

Q You said that the frequency of genetic diseases has increased—that in 1900 there were 5 deaths from genetic disorders for each 1,000 live births. In 1980, there were still 5 genetic disorder deaths for each 1,000 live births. I do not see what you mean by an increase in genetic disorders.

A There is no *absolute* increase, but there is a *relative* increase. At the beginning of the century, the infant mortality was 150:1,000 live births. Five of these 150 deaths were related to genetic causes. The remaining 145 were due to nongenetic causes, such as infections. In 1980 there were 15 infant deaths for each 1,000 live births. The same *number* of deaths among those (5) is still due to genetic causes or partly genetic causes. Apparently, the other 10 in every 1,000 births in 1980 were due to other causes, e.g., poor infant nutrition, infectious diseases, or accidents. These data show that there is no epidemic of genetic disorders but that *relative* frequency and *relative* importance have increased.

Q Doctor, if I heard you correctly, you mentioned that many times chromosomal disorders lead to spontaneous abortions. Why is that?

A Yes. I think Chapter 2 may discuss more about this subject because it is extremely interesting and because physicians have learned a great deal about it recently. The data from more then the 60,000 consecutive newborns who underwent chromosome analysis show that the frequency of chromosomal disorders in the newborn period is 0.6 percent. There is some data from spontaneous abortion materials, especially in the early phase of pregnancy. The number might be different in very early pregnancy and a bit later, but overall it is almost 60 percent. This is the nature of selection, in a way; the fetuses are too seriously affected to come to term, and they are thus eliminated by spontaneous abortion.

Q Doctor, this is an ethically loaded question, but with no malice aforethought, I just wondered, when you mention the Meckel-Gruber syndrome, prenatal diagnosis, and prevention of the defect, are we really speaking about preventing the *birth of the child* with the defect or the *defect* itself?

A I was talking about preventing the birth of the defective child. Recent intrauterine surgery on fetuses with hydrocephaly and hydronephrosis, however, indicates that certain defects may be partially corrected in utero.

Q I notice you did not say anything about smoking, and I know that quite a bit of the literature now speaks about that, nor did you say anything about the father's contribution to fetal abnormalities.

A Smoking in itself, as far as is known, does not cause birth defects, but most likely, because of vascular compromise during pregnancy, it causes low birth weights in the newborn babies. Fathers come to physicians' attention indirectly because of the incidence rates in the chromosomal abnormalities. Society hears a lot about the pregnant woman's age (e.g., under or over 35) but not about the father's age. Recently, some data, mainly from Europe, suggests that the father's age may also be important

11

in chromosomal abnormalities if he is about age 50 or 55 (or, most recently, above age 40). A 1981 article discusses this subject (*Human Genetics* 59 [1981]:119). Also, in this group of fathers of advanced age, fresh mutations apparently are increased; these are mainly dominant disorders. Where the previous generation is healthy, and the father and mother are genetically healthy but the father is old, there is an increased risk for mutational disorders in the couple's offspring. A 1975 article examines this topic (*Journal of Pediatrics* 86 [1975]: 84-88).

Q Doctor, you mentioned the relationship of alcohol intake by the mother to some of the disorders. What about the alcoholic father? Is there any association there?

A I am glad you reminded me. The problem is that the father's contribution and the mother's contribution to the fetus are a bit different because the cytoplasm of the sperm is very small relative to the amount of cytoplasma of the human ovum. Some recent papers suggested, however, that toxic substances may even be transmitted from the father to the fetus. One of these toxic substances might be alcohol. There is some data for the paternal fetal alcohol syndrome. As far as I know, however, it has not yet been substantiated adequately.

Q Dr. Taysi, into which class of genetic disorders would phenylketonuria (PKU) fall?

A The single-gene-disorder group. Patients with PKU have an enzyme deficiency (phenylalanine hydroxylase) due to a single abnormal gene. I did not have time to go into detail, but it takes two genes with the same mutation, one from each parent (when both parents are carriers), for a disease from this group to manifest itself in the offspring.

Q In your presentation, you mentioned situations where there is mental retardation as well as physical abnormality. What data are there on mental diseases, for instance, schizophrenia? Are not some mental diseases associated with a genetic disorder?

A That is a very interesting subject. The data, however, are not conclusive. Schizophrenia and some other mental disorders, e.g., the manic-depressive syndrome, might be due to multifactorial inheritance. Some data show the recurrence risk in families with schizophrenia to be roughly 10 percent. Regardless of whether a child is living with a schizophrenic parent or not, there is a 10 percent recurrence rate. Apparently, there is some genetic component, even if it is not well defined.

Footnotes

1. M. M. Kaback, "Medical Genetics: An Overview," *Pediatric Clinics of North America* 25 (1978): 395-409.

2. H. Galjaard, "Congenital Disease and (Infant) Mortality and Morbidity," in *Genetic Metabolic Diseases,* H. Galjaard, ed. (Amsterdam: Elsevier-North Holland Biomedical Press, 1980), p. 6.
3. J. G. Hall, E. K. Powers, R. T. McIlvaine, and V. H. Ean, "The Frequency and Financial Burden of Genetic Disease in a Pediatric Hospital," *American Journal of Medical Genetics* 1 (1978): 417-436.
4. B. K. Trimble and M. E. Smith, "The Incidence of Genetic Disease and the Impact on Man of an Altered Mutation Rate," *Canadian Journal of Genetics* 19 (1977): 375-385.
5. V. A. McKusick, "Mendelian Inheritance in Man," in *Catalog of Autosomal Dominant, Autosomal Recessive and X-linked Phenotypes,* 5th ed. (Baltimore: Johns Hopkins University Press, 1978).
6. D. W. Smith, *Recognizable Patterns of Human Malformations,* 3rd ed. (Philadelphia: W. B. Saunders Co., 1982), p. 98.
7. Smith, p. 140.
8. Smith, p. 18.
9. J. E. Seegmiller, "Inherited Deficiency of HGPRT in X-linked Uric Aciduria (The Lesch-Nyhan Syndrome and Its Variants)," *Advances in Human Genetics* 6 (1976): 75-163.
10. Smith, p. 248.
11. Smith, p. 10.
12. C. E. Anderson, J. I. Rotter, and J. Zonana, "Hereditary Considerations in Common Disorders," *Pediatric Clinics of North America* 25 (1978): 539-556.
13. J. E. Kurent and J. L. Sever, "Infectious Diseases," in *Handbook of Teratology,* 2nd ed., vol.1, J. G. Wilson and F. C. Fraser, eds. (New York: Plenum Press, 1979), p. 225.
14. Kurent and Sever, p. 169.
15. Kurent and Sever, p. 316.
16. Kurent and Sever, p. 398.

• chapter 2 •

Genetic Diagnosis

Rev. Albert S. Moraczewski, OP, PhD
(Redactor)

Impact of Genetic Disease

This presentation will begin with an initial statement on the impact of genetic disease in U.S. society, which reinforces what Dr. Taysi has stated in Chapter 1. It is of interest that genetic diseases play a major role in the health care concerns of the United States. For example, there are about 15 million Americans with some sort of birth defect, 80 percent of these of genetic origin. About 60 percent of early miscarriages, i.e., first-trimester spontaneous abortions, are due to a chromosome abnormality, which physicians now know to be a major cause of early reproductive loss. About 4 percent of liveborn infants have some sort of congenital malformation, and about 1:200 infants (0.5 percent) have a chromosome abnormality. Approximately 40 percent of infant mortality is related to genetic factors and diseases, and about half of all new cases of mental retardation occurring each year are related to genetic disorders. In sum, Blue Cross/Blue Shield estimates that around 20 percent of total U.S. health care costs are related to genetic diseases.

A question frequently asked is, "Do genetic diseases occur in adults and account for a major part of hospitalization days in adults?" The answer is affirmative, in that physicians are learning that an increasing number of common, familiar diseases such as hypertension, arterial chronic heart disease, and mental illness (schizophrenia and manic-depressive disorders) do have

---

Fr. Moraczewski is vice president for research, Pope John XXIII Medical-Moral Research and Education Center, St. Louis, MO. This chapter is based on the presentation of James P. Crane, MD, at the Genetics and Health Care Seminar, October 6, 1981.

some familial component. There also may be some environmental factors required for expression of the disorder, but a person's genetic predisposition certainly plays a role. Clearly, to some extent genetic diseases do play a significant role in adults.

The role of chromosomal abnormalities in bringing about miscarriages was mentioned above in passing but needs additional comments. Of 1,000 conceptions, about 850 will result in live births and about 150 will result in miscarriage. Miscarriage is thus quite common, occurring in approximately 15 to 20 percent of all recognized pregnancies. The word *recognized* is emphasized because there are now more sophisticated tests for earlier diagnosis of pregnancy. Miscarriage may have been even more common but often goes undiagnosed or unrecognized because it occurs so early that the woman interprets it as a menstrual period that is a few days or weeks late.

Of the 850 live births, the vast majority of those infants will be chromosomally normal, while approximately 4 will have some chromosome abnormalities. Of the 150 miscarriages, 60 percent of which are chromosomally abnormal, about 90 chromosome abnormalities would have occurred if the pregnancies had come to term. Hence, about 94 of 1,000 conceptions are chromosomally abnormal—roughly 1:10. Chromosome abnormalities are a relatively common occurrence. If we are to believe some of the newer published data on early unrecognized miscarriages, it appears that perhaps as many as one third to one half of all conceptions may be chromosomally abnormal and end in miscarriage. This observation provides some idea of the frequency of various chromosome abnormalities in liveborns as compared to all conceptions. For example, 1 of every 118 conceptions has an extra no. 21 chromosome (trisomy 21, Down syndrome). Yet, the frequency of trisomy 21 among live births is about 1:900. Hence, the survival rate of fetuses with trisomy 21 from conception to birth is about 1:8. There are other very common chromosome abnormalities, e.g., trisomy 16, which occurs once in every 74 conceptions but is never found in liveborns. Because it is a fatal abnormality, the survival rate is zero.

Prenatal Diagnostic Tools

Among the various prenatal diagnosis tools, there are four current ones of which three will be briefly discussed, and one at some length: ultrasonography, amniography, fetoscopy, and amniocentesis. *Ultrasound* (ultrasonography) involves the use of high-frequency sound waves that are above the range of human hearing. These sound waves can be transferred through the pregnant uterus and will be reflected back by any structure they strike.

This reflected wave, or echo, can then be amplified by an appropriate amplifier-computer circuit and displayed on a television screen to make an image. With the use of ultrasound, one can actually visualize the fetus and even structures within the fetus.

A number of congenital malformations can be diagnosed with ultrasonography, especially where a part of the fetus is absent or of inappropriate size or location. Anencephaly, one of the neural tube defects, can be so diagnosed. Where one would expect to see a circular outline in the ultrasonogram, there only a few segments indicating the base of the skull and facial features. That view later can be correlated with the actual fetus in order to compare scan findings with reality.

Another example of a neural tube defect detectable by ultrasound is meningomyelocele. In an ultrasonogram through the fetal pelvis, the thigh, knee, calf, and fetal bladder are visible. The meningomyelocele will be seen as a large cystic structure protruding off the back of the spine. Often such a case of a neural tube defect actually involves multiple birth defects. The fetal liver and intestines are outside the abdomen, while the heart is outside the chest. In the condition called holoprosencephaly, an ultrasonogram of the fetal head will appear empty because of the absence of structures within it. It can occur in conjunction with certain chromosomal abnormalities, e.g., trisomy 13 or trisomy 18. It also may occur as a sporadic birth defect. Holoprosencephaly is nonhereditary, although it seems that also in certain families it occurs as a recessive trait. It is fatal. Very often, infants afflicted with this condition will have facial malformations such as narrowly spaced eyes and a single nostril. In its severest form, this condition is called cyclopia, a deformity where there may be only a single fused eye in the midline, with a proboscis or tubular appendage protruding above the eye.

Another example of a congenital malformation, an instance of which recently made news in California, is called type 4 Potter syndrome. In this malformation the urethra lacks an opening so that the baby is unable to void, or there are no posterior urethral valves so that the baby cannot urinate. The abdomen thus becomes markedly distended as fluid accumulates in the urinary bladder. The bladder wall becomes thinned out to the point that urine transudes into the abdominal cavity.

A recent report documented in utero treatment of this condition by placing one end of a catheter into the fetal bladder and the other end into the amniotic sac, which permitted drainage of the abnormal fluid accumulation. The baby survived.

Finally, dwarfism is often diagnosable through ultrasound. Because the chest cage is underdeveloped in this condition, the abdomen appears to protrude. The feet, toes, arms, and legs are markedly shortened.

Amniography, another prenatal diagnostic tool, is becoming less commonly used. This technique involves inserting a needle into the amniotic sac that surrounds the fetus and injecting a radiopaque dye. The dye will outline the fetus and permit the taking of an x-ray image that, although not containing enough radiation to harm the fetus, is sufficient for diagnostic purposes. It can provide enough information to diagnose certain birth defects. Amniography has been largely replaced by ultrasonography, however, and there are very few instances in which amniography is the procedure of choice.

Fetoscopy refers to an increasingly used technique for prenatal diagnosis. Its ultimate role, however, will more likely be in fetal therapy and intrauterine surgery. This technique involves the use of a fibroptic light system and a needle, or small cannula, about 2 or 3 mm in diameter. The fetoscope allows the physician instrumentally to enter the uterus and directly visualize the fetus and to determine whether any abnormalities are present. For example, the physician can sample blood from the fetus by obtaining blood from the placenta or can take biopsies of fetal skin to diagnosis certain genetic disorders. Some clinicians believe that fetoscopy will become increasingly used and find its major role, however, in the in utero treatment of congenital malformations.

At present, however, fetoscopy should be considered a research tool. The mortality, or pregnancy loss rate, associated with the procedure normally runs between 3 and 6 percent. Hence, there is a significant risk of losing the pregnancy just by doing the procedure. But there is hope for the future in some exciting work being done at the pregnancy research center in Bethesda, MD, in animals (primarily monkeys). Fetoscopy is being used to treat neural tube defects, specifically hydrocephalus. Under fetoscopic guidance, a device is inserted through the fetal skull and into the ventricle in order to drain the excess fluid while the fetus is still in utero. The researchers report that they have been quite successful in preventing and treating hydrocephalus (in animals) with this technique. The same research group is also working on the development of a paste material containing bone particles that might be useful in treating spina bifida in utero. The pastelike material is applied to the defect and allows bone to grow and cover the defect.

Although definitive results are a long way away, researchers anticipate that more and more treatments will be developed to correct genetic defects in utero.

The last prenatal diagnostic technique to be discussed here is amniocentesis, which is probably what most people think of when they think of prenatal diagnosis. The remainder of this chapter will examine amniocentesis: the technique, the indications, and the risks.

Amniocentesis

The Technique. Amniocentesis is generally done at about 15 weeks after the last normal menstrual period—about 13 weeks after conception. The amnionic fluid volume is about 150 ml at that time. A needle is gently inserted through the abdominal wall into the amnionic cavity, and 20 to 30 ml of amnionic fluid is removed. Removal of the fluid will not adversely affect mother or fetus, since the amnionic fluid volume will regenerate itself over a few hours. The amnionic fluid obtained contains desquamated cells that have fallen from the surface of the fetal skin; some of these cells are still alive. These cells are then grown in a special culture media. Once an adequate quantity of cells is grown—in 3 to 4 weeks—the cells can be harvested. The cells' growth is stopped by the chemical substance colchicine, and by additional processing the chromosomes are made visible under the microscope. A karyotype is made by taking photographs of the specially stained chromosomes through high-power magnification ($1000\times$). The photographs of the chromosomes are cut out and arranged in a standard pattern for analysis and the determination of the presence or absence of certain chromosomal abnormalities.

Indications. The first of six major indications for amniocentesis is *advanced maternal age.* In most institutions and generally across the United States, "advanced" maternal age means age 35. It has been selected by medical geneticists as an age beyond which parents should at least be advised of the fact that there is an increasing frequency of chromosome abnormalities. Most clinics will offer these couples the option of prenatal diagnosis through amniocentesis. Clearly, there is nothing magic about age 35; the body does not suddenly fall apart at 35 while being perfectly intact at 34. It is an arbitrary cutoff point. The risk of a chromosomally abnormal conception at age 35 is about 1:400, while at age 34 it is 1:500. It is primarily a matter of selecting a cutoff point beyond which the procedure is recommended. There are a variety of reasons for the selection of this particular point; suffice it to say here that most health care facilities use 35 as the cutoff age (see Figure 1). After age 35 there is a relatively steep rise in the frequency of chromosomal abnormalities; by age 45, the risk is approximately 10 percent, or 1:10 conceptions.

Although not all chromosome abnormalities are associated with advancing age, most are. Although either the mother or the father may contribute an extra chromosome to cases like trisomy 13, trisomy 18, and trisomy 21, about half of all cases will occur in mothers who are aged 35 or over, even though they represent only about 7 percent of the reproducing population. These chromosome abnormalities are not limited to older mothers, but they are certainly more common as maternal age increases. The severity of the

DOWN SYNDROME

FIGURE 1.—The relationship between maternal age and the likelihood of trisomy 21. The sharp increase in the risk for trisomy 21 occurs at 35 years of age; this is the age currently chosen as an indication for prenatal diagnosis.

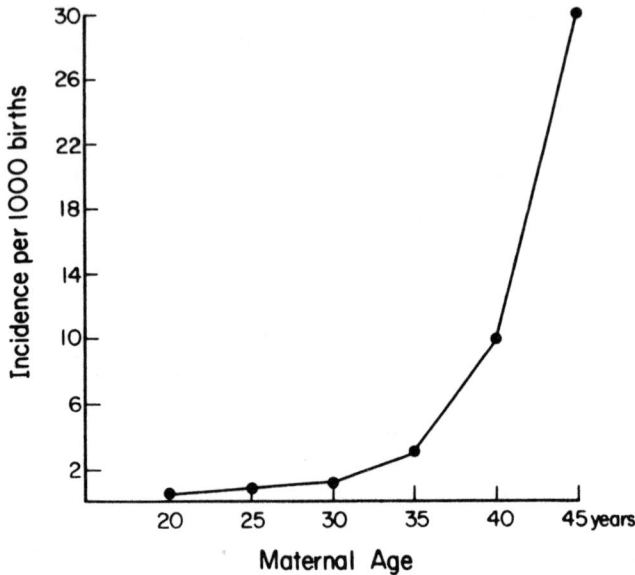

SOURCE: Adapted and reproduced with permission from T. E. Kelly, *Clinical Genetics and Genetic Counseling.* Copyright © 1980 by Year Book Medical Publishers, Inc., Chicago.

defects varies. In Down syndrome (trisomy 21), for example, if an afflicted child does not have congenital heart disease, he or she has a reasonable life expectancy of about 55. On the other hand, trisomy 13 is a fatal condition in that about half of affected infants will die by the age of 1 month. The longest recorded survival is about 10 years, because infants with trisomy 13 are severely handicapped. They tend more frequently to have congenital heart disease or to be much more severely mentally retarded and are characterized by microcephaly and other physical handicaps.

The second major indication for amniocentesis is *a couple who have had a previous child with a nondisjunctional chromosome abnormality,* in which chromosomes in either the egg or the sperm divided unequally so that the child has either one too many or one too few chromosomes. The classic example of this condition is trisomy 21, or Down syndrome (described above), in which parents who have had a previously affected child have about an 0.5 to 1 percent risk of some type of recurrent chromosome abnormality in each subsequent pregnancy. It does not necessarily have to be Down syndrome; in other words, one child could have Down syndrome and the next child could have trisomy 13 or trisomy 18. The basic problem is that there is a predisposition for the nondisjunctional phenomenon to occur. Overall, about two thirds of the cases of Down syndrome are maternal in origin, that is, the mother contributes the extra chromosome.

A third recognized indication for offering amniocentesis is when *a parent is a carrier of a chromosome abnormality*. In this situation the child's chromosomal abnormality does not occur by chance and is not related to age and maldivision of the chromosomes in either the egg or the sperm. Rather each cell in a parent's body already has a chromosome abnormality, or what is called a *chromosome translocation*. A few cases will illustrate this condition.

The karyotype of some children with Down syndrome will show that there are apparently only 2 no. 21 chromosomes, i.e., the normal number, one contributed by the mother and the other contributed by the father. The no. 4 chromosome pair, however, is unequal in length: one chromosome of the pair is a little longer than the other. This is due to the fact that the longer one has, attached to its upper end, a third no. 21 chromosome; this condition is thus called a 4/21 translocation. About 5 percent of children with Down syndrome have an unbalanced translocation state of this sort. By merely looking at such children, it is impossible to distinguish them from children who have the extra free no. 21 chromosome, i.e., who have trisomy 21. It is important to do a karyotype of children with Down syndrome because the parents can then receive accurate counseling on the risk of recurrence. If the child has trisomy 21, the risk is only 0.5 to 1 percent. On the other hand, if the problem is an unbalanced translocation state, it is conceivable that one of the parents may be a carrier for that translocation.

For example, the karyotype of such a child's father might show that there is only one free no. 21 chromosome, while the second one is attached to one of the no. 4 chromosomes. This condition is called a balanced translocation state. In other words, there is a correct amount of genetic information contained in the pair of no. 21 chromosomes, but they are structurally rearranged. Because there is a correct amount of genetic information in the father, he will be perfectly normal. The problem arises when he reproduces, because he will then pass only one chromosome of each pair of his chromosomes on to his children. If he passes his normal no. 4 and his normal no. 21 to his child (and his wife does the same), they will have a normal child. If he passes his normal no. 21 and the no. 4 with the attached no. 21 (and the mother contributes a normal no. 4 and no. 21), however, the child will get one too many chromosomes (i.e., the unbalanced state) and will be affected. The risk of recurrence is much higher in this situation. If the father is the carrier, as in this particular example, the risk of recurrent Down syndrome in a pregnancy is about 3 percent. If the mother is the carrier, however, the risk is about 10 percent. The reason for the resulting difference in recurrence risk between paternal and maternal carriers is not understood. That is not true of all translocation conditions, but it is true in this particular situation.

Two more examples of an unbalanced translocation state will serve to illustrate other aspects. In one particular case, for example, there are 3 no. 21 chromosomes, two of them attached end to end. This is a 21/21 unbalanced translocation. One parent's karyotype would show 2 combined no. 21 chromosomes, while the single no. 21 would be absent. Two things can happen. One, during reproduction the parent will not pass on a no. 21 chromosome and miscarriage will occur. (This is a lethal abnormality, and there has never been a liveborn child with this condition.) Two, the parent will pass on both no. 21s fused together, and when the other parent adds his or her no. 21, the child will end up with an unbalanced chromosome arrangement. In this type of chromosomal unbalance there is really no way in which the parents can have normal children; the only possibilities are miscarriages or children with Down syndrome.

The final example of a translocation state involves two brothers, both severely mentally retarded. Their intelligence quotients (IQs) are in the range of 30 to 35. They also have somewhat peculiar facial appearances—the eyes appear deepset, the chin is prominent, and the nose is a little rounded and bulbous. In this particular case the mother happens to be the carrier. Her karyotype reveals that the translocation involves one of the no. 4 chromosomes and one of the no. 9s. There is a piece missing off the end of one no. 4 chromosome, and that missing piece is attached to the top of one of the 9 chromosomes. The mother has a correct amount of genetic information, and so is normal, but the children received her normal no. 4 and her no. 9 with the extra segment of no. 4 material. The father contributed a normal no. 4 and no. 9, with the result that the children have extra chromosome material and are handicapped, primarily mentally in this particular case. The risk of recurrence with this particular translocation is around 10 percent.

Parents who are carriers of inborn metabolic errors, are a fourth indication for offering amniocentesis. These are diseases in which generally there are enzyme deficiencies. The vast majority of these fall into the autosomal recessive category of inheritance, where both parents are carriers and there is a 1:4 risk of recurrence for each pregnancy if there is a previously affected child.

TABLE 1. Selected List of Conditions Eligible for Prenatal Diagnosis by Biochemical Analysis of Cells from Amniotic Fluid

Disease	Disease
Acid phosphatase deficiency*	Krabbe disease*
Adenosine deaminase deficiency*	Lesch-Nyhan disease*
Argininemia	Maple syrup urine disease*
Argininosuccinic aciduria*	Menke disease
Citrullinemia	Metachromatic leukodystrophy*
Cystathioninuria	Methylmalonic aciduria*
Cystinosis*	Mucolipidosis II (I-cell disease)*
Fabry disease*	Mucopolysaccharidosis I (Hurler)*
Fucosidosis	Mucopolysaccharidosis II (Hunter)*
Galactokinase deficiency	Mucopolysaccharidosis III*
Galactosemia*	Mucopolysaccharidosis IV*
Gaucher disease*	Niemann-Pick disease
Glycogenosis II*	Ornithine carbamyl transferase deficiency
Glycogenosis III	Phenylketonuria, "lethal" type
Glycogenosis IV*	Porphyria, acute intermittent
GM_1-gangliosidosis*	Propionic acidemia*
GM_2-gangliosidosis (Sandhoff)*	Sickle cell anemia
GM_2-gangliosidosis (Tay-Sachs)*	Thalassemias
Homocystinuria	Wolman disease*
Hyperammonemia II	Xeroderma pigmentosum*
Hypercholesterolemia, hereditary	

*Diagnosis has been done.

SOURCE: Reproduced from J. J. Nora and F. C. Fraser, Medical Genetics: Principles and Practice. Copyright © 1981 by Lea & Febiger, Philadelphia, PA. Printed with permission.

Table 1 lists some of the conditions that can be diagnosed prenatally. A few examples will have to suffice for discussion. The first is a condition called I-cell, or Leroy syndrome; it is one of the mucolipidoses. These disorders are due to deficiency in a number of lysosomal enzymes, which causes severe mental and growth retardation. Also associated with the condition are a relatively coarse facial appearance (which is characteristic of many of these types of diseases), a failure to thrive, and ultimately death due to either congestive heart failure or pneumonia, generally at 2 to 3 years of age. Both parents must be carriers for the disorder to occur. The risk of recurrence in subsequent pregnancies is 1:4. With amniocentesis the condition can be prediagnosed prenatally by measuring these enzymes in fetal skin cells that have been cultured from the amnionic fluid. A second example in this category of inborn metabolic errors is called methylmalonic acidemia. It is a recessive trait in which both parents are carriers, with a 1:4 risk of recurrence. The symptoms of the disease include mental retardation, growth deficiency, recurrent episodes of nausea and vomiting, acidosis, coma, and death in early infancy. The condition can be diagnosed through prenatal analysis of cultured amniotic fluid cells. It is one of the few conditions that is currently treatable: certain forms of this condition are responsive to vitamin B_{12} when it is given in massive doses to the pregnant mother. An adverse outcome in the infant can thus be prevented.

The problem with most of these autosomal recessive diseases is that physicians never see a couple at risk until they have had their first defective child, because nothing in their family history warns them that they are carriers for a disorder. There are a few exceptions, however, and one is Tay-Sachs disease. Although it is a recessive trait, physicians can screen couples at risk, since the disease is more common in Jewish persons of Eastern European extraction. About 1:30 of this ethnic origin will be a carrier for the condition, which can be determined by a simple blood test. If both parents are carriers, they can be offered a number of reproductive options: electing not to have children, prenatal diagnosis of the disorder, artificial insemination with a *donor* sperm (which would eliminate the father as a carrier). This devastating disease is always fatal and is marked by progressive neurological deterioration, muscle weakness, blindness, deafness, seizures leading to a vegetative state, and death.

The fifth category of indications for amniocentesis is the one that includes *mothers who are carriers for X-linked diseases,* of which 109 are now recognized. The major recognized X-linked disorders most frequently seen are muscular dystrophy (Duchenne's disease) and hemophilia, although prenatal diagnosis can be used in a number of other X-linked diseases. These diseases are carried by the mother. She is usually not affected by the disease, but her sons have a 50 percent chance of being affected. Her daughters are not at risk for the disease, although half of them will be carriers for the disorder, like their mother.

Another example of an X-linked disorder is the Lesch-Nyhan syndrome. It is one of three X-linked diseases that can currently be diagnosed through amniocentesis. Others can be diagnosed with ultrasonography, for example. The Lesch-Nyhan syndrome is characterized by an enzyme deficiency, which results in the variety of problems discussed in Chapter 1.

One of the problems with X-linked diseases is that unfortunately many of them cannot yet be diagnosed in utero, so that many times physicians can tell couples only that they will have a son who has a 50 percent risk of being affected or daughter who is not at risk, but they may not be able to say specifically that the son will be affected.

The sixth indication for amniocentesis is *a family history of neural tube defects*. This includes conditions such as anencephaly (partial absence of the brain), a fatal condition that occurs in about 1:1,000 births; encephalocele (protrusion of brain tissue through an aperture in the skull), a related defect that generally is fatal; and meningomyelocele (protrusion of the spinal cord through a defect in the vertebrae), which may not be fatal, depending on the size and location of the defect, whether it is open or closed, and what sort of complications occur during the neonatal period.

Collectively, these malformations affect about 1:500 pregnancies in the general population. Generally, the family history for similar abnormalities is negative. In 90 percent of the cases, there has not been a previously affected child. On the other hand, if parents previously have had an affected child, there is a 2 to 3 percent risk of recurrence in subsequent pregnancies. Similarly, a person who has spina bifida has a 2 to 3 percent risk of having an affected child. If a couple has had 2 affected children, the risk for the third child increases to about 5 percent. Normal siblings of an affected child have about a 1 percent risk of having a child with a neural tube defect when they reproduce. That risk is above the risk in the general population.

Prenatal diagnosis in these cases is basically twofold. The first is amniocentesis, followed by measurement of alpha-fetoprotein in the amnionic fluid. Alpha-fetoprotein is a rather unique protein produced by the fetal liver that decreases in concentration as the pregnancy nears term. Normally, only small quantities of alpha-fetoprotein are found in the amnionic fluid. If the fetus is affected with an open neural tube defect, i.e., not a skin-covered lesion, however, an excess quantity of alpha-fetoprotein, greater than three standard deviations, will be found in the amnionic fluid, and this allows detection of such abnormalities. The false-negative rating in this measurement is about 10 percent; in other words, about 10 percent of cases will be missed with the alpha-fetoprotein test apparently because the lesions are skin covered and alpha-fetoprotein leakage does not take place or is greatly reduced. The false-positive rating is quite low, about 1:1,000, and has decreased as additional ancillary tests have become available to double check abnormal results. One clinic, combining alpha-fetoprotein and ultrasonography studies, has attained 100 percent accuracy in diagnosing neural tube defects over the past 5 years.

Risks. There are basically three things to be considered concerning the risks of amniocentesis. *First,* it was originally thought that amniocentesis could induce miscarriage. Geneticists were quite concerned about this at one time. With more recent data, however, there is now a body of evidence to suggest that the incidence of miscarriages in women undergoing amniocentesis is not statistically different than the incidence of miscarriages in the general population, at least when done under ideal conditions.

The *second* concern is fetal trauma. Could the needle actually enter the baby and cause some damage or birth defect? Because the procedure deals with a relatively small space, this is a legitimate concern. The technique of amniocentesis has been honed, however, and the procedure is now quite a bit safer. In many clinics, all the procedures are done under direct ultrasound guidance so that the needle is actually visualized as it enters the uterine cavity and the fetus is kept in constant view. Any movement of the fetus is apparent, and care is taken to keep the needle as far away from the fetus as possible. Moreover, the needle is inserted as far away as possible from the fetus and the placenta because one of the other complications of the procedure is fetal-placental hemorrhage. In the United States as a whole it has been reported only a few times. Generally, fetal-placental hemorrhage can be avoided by localizing the placenta with ultrasound and selecting a site away from both the placenta and the fetus.

The *third* potential complication of the procedure is amnionitis, i.e., an infection developing within the uterine cavity. This appears to be very rare. In one institution, over 10,000 amnioceneses were performed without one case of amnionitis. It has been reported as an isolated event in other institutions, hence it must be recognized as another potential risk. Amnionitis can be very serious because it can cause uterine irritability, premature labor, and loss of the fetus. In addition, it can cause an overwhelming infection in the mother, possibly giving her subsequent infertility problems. Even though amnionitis is rare, it can be very serious. It can be prevented by care and the use of a sterile technique.

As our knowledge of genetic disorders increases and diagnostic techniques have improved in accuracy and safety—for mother and child—there is a growing need for increased research in the treatment of genetic related conditions. Such treatment will be directed not merely to the amelioration of symptoms but more radically to the correction of the basic genetic defects.

Discussion*

Q You mentioned, Doctor, some of the possible negative effects in connection with amniocentesis and cited recent data showing that there was no difference between the rate of spontaneous abortion resulting

---

*The printed responses are based on the speaker's comments and are not a verbatim report.

from amniocentesis as compared to the general population of pregnant women. Data published about 3 years ago, which included North American and European studies, indicate a somewhat higher rate, about an additional 1.5 percent. How do you account for the differences?

A The background frequency of miscarriage in the second trimester is about 1:50 women (2 percent). In other words, the mother who is 15 or 16 weeks pregnant has about a 1:50 risk of miscarrying. In doing these sorts of studies, it is very important to know if appropriate controls were used. For example, the frequency of miscarriage increases with advancing maternal age, so it is not valid to compare a group of mothers aged 35 and over with a group of mothers aged 20, because one would expect the older mothers to have an inherently higher rate of miscarriage. The two major studies in this continent have been a Canadian study and the collaborative study by the National Institute of Child Health and Development. Both showed, in comparing controls versus mothers undergoing the procedure, that there was no difference in the frequency of miscarriage. Furthermore, it is possible to have actually a lower rate of miscarriage in mothers undergoing the test than were reported in either of those two studies. This is not to say that the procedure prevents miscarriage, but what happens, may be explained by the following: Because about 1 in every 10 mothers who seeks counsel and elects to undergo amniocentesis is 40 years of age or older, it is not uncommon to find, when preparatory ultrasound examination is carried out, that the mother is not carrying a viable pregnancy. An amniocentesis is not performed on such patients. The patient subsequently will miscarry. Had amniocentesis been done without the preparatory ultrasound examination, the miscarriage could have attributed to the procedure. Because in these instances such women have been eliminated from the patients on whom amniocentesis is done, the calculated miscarriage rate is accordingly quite low. One British study suggested a higher rate of miscarriage in mothers undergoing amniocentesis. It is important to understand the reasons why women in that particular group underwent the procedure, however. The majority of them did so because they had elevated serum alpha-fetoprotein levels. In other words, during a screening process, a blood test revealed an elevated alpha-fetoprotein level, and amniocentesis was done to confirm the presence or absence of a neural tube defect. The problem of alpha-fetoprotein elevation is nonspecific, since a number of conditions will have that effect, one of which is impending miscarriage. Hence, if amniocentesis is done on pregnant women who have elevated serum alpha-fetoprotein, that group is at higher risk for miscarriage than is the general population. That phenomenon may explain the discrepancy in findings between the studies.

Q Doctor, you mentioned there was some chromosome abnormality at the age of 35. I believe you quoted the rate of 1:400 at the age of 35. What are the odds when the mother is younger?

A At age 33 the rate is about 1:600; at 32 years, 1:700; 31 years, 1:800; 30 years, 1:900; whereas, at age 25 it is about 1:1,200.

Q Since the chromosome makeup of the parents does not change during those years, are the changes in odds due to environmental factors?

A Probably the best theory to explain the female age-related phenomenon is related to the difference in the way egg cells and sperm cells are formed. A new supply of sperm cells is generated about every 72 days, but the woman, even before birth, possesses all the eggs she will ever have. These eggs begin dividing before birth but then stop and do not resume dividing until ovulation, which, for example, in a pregnant woman of 35, would be 35 years after the initial steps of division. It is felt that this long period of delay between the starting of the division process and its completion may increase the probability of something going awry. That could account for the higher frequency of chromosomal abnormalities with increasing maternal age. That is just one theory; there are several others, but time is lacking to discuss these.

Q About 10 years ago or so, we learned in medical school that the pregnant teenager experiences an increased incidence of trisomy 21. Have any new data changed this view today? In addition, is there, in general, an increased incidence of hereditary problems in women with metabolic disorders such as diabetes?

A There does appear to be a slightly increased risk of chromosome abnormalities in very young (teenaged) mothers. It is not as dramatic an increase as one sees with advancing maternal age. About diabetes, there is definitely a higher incidence of birth defects in mothers who have diabetes. In fact, the frequency of birth defects is related to the severity of the mother's diabetes. For example, if the mother has advanced diabetes with vascular disease, there is about a 16 or 17 percent incidence of major malformation. Diabetic women should be counseled about this particular risk before attempting a pregnancy. It is a tragic situation when such a mother undertakes a pregnancy, very rigidly controls her diabetes, does everything she is supposed to do, and ends up with a severely affected child.

Q I have one question in regard to fetoscopy. As I recall, there is a greater incidence of trauma associated with fetoscopy than with amniocentesis. Is the difference merely due to the greater diameter of the catheter or needle being inserted, or do other factors account for the difference in

the frequency of trauma?
A There are several factors. Differences in needle diameter may be one, but probably one of the major factors has to do with cases where fetal blood sampling had been done. In that category, there have been reports of fetal exsanguination. The procedure involves visualizing a placental blood vessel, inserting a needle, and drawing blood, just as one would draw blood from a vein. Most of the time the bleeding will top spontaneously, but there have been reports of fetal death resulting from uncontrolled hemorrhaging. There is no question that the procedure is hazardous.
Q But if the procedure did not involve actual blood sampling or a biopsy of the fetus, but rather an attempt to visualize the external appearances of the fetus, would there then be a difference between such fetoscopy and amniocentesis?
A There would still be a difference, although it is likely that the major portion of the morbidity associated with the procedure is probably related to bleeding problems. Whether the morbidity of fetoscopy, when done for blood sampling purposes, has been compared to the procedure when done for just visual inspection is problematical.

chapter 3

Genetic Counseling

Rev. Robert Baumiller, SJ, PhD

    This chapter will attempt to define what genetic counseling is and how it is accomplished. Beyond this goal, I will try to share both experiences in the field and the situations that my rather unique combination of roles (geneticist and priest) has presented.

    Genetic counseling is an unfortunate title given to a process of genetic information giving. It is unfortunate because it leads some clients to expect guidance in their ultimate decisions about present or subsequent reproduction. It is unfortunate also because such client expectations (and, indeed, the term *counseling* itself) move some counselors to provide such guidance. Genetic counseling may be carried out by a physician, a medical geneticist with a doctorate, or a genetic associate with a master's degree in science plus a degree in either biology, nursing, or social work. The process is generally carried out with one client, one couple, or one family group, since any particular genetic disease is rare and its effect family centered. Today, in certain circumstances, genetic counseling is carried out with larger groups. For example I see 12 couples at a time when a common nonfamilial indicator of increased risk is present. The most frequent use of this type of counseling is for advanced-maternal-age risks, but community exposure to mutagens might also necessitate larger group counseling.

    When a physician does genetic counseling, it is most often in conjunction with diagnosis or ongoing treatment and most frequently involves the affected person alone or with his or her parents if the affected person is a minor. Many physicians are familiar with the basic principles of genetics, but

---

Fr. Baumiller is director, Division of Genetics, Department of OB/GYN, Georgetown University Hospital, Washington, DC. The editorial suggestions of Anthony Waddell are appreciated.

in this rapidly advancing field only a very specialized group—the clinical geneticists—are sufficiently trained and knowledgeable to provide the comprehensive information needed. Clinical geneticists do become involved with the families they care for and supply an important part of counseling. Even in, or perhaps especially in, genetic clinics, however, counseling is recognized as a separate function and is most frequently carried out by a medical geneticist or a genetic associate.

From more than 20 years of experience, I will outline my perception of genetic counseling. My outline may not represent what is done universally, but there are broad areas of agreement. I would like to think that at Georgetown University we are person oriented and enter into the decision-making process more deeply than most divisions of genetics. If my perception is true, it might be due to our particular bias about human values and human freedom.

I will describe first genetic counseling for specific diseases and then what we call preamniocentesis counseling. In my office, counseling is carried out by myself or by a physician who has finished her obstetrics-gynecology training and is now, as an instructor, learning genetics as a subspecialty. Counseling may also be done by a medical geneticist who trained with us and who is a Carmelite priest. We also have a genetic associate available for counseling. None of us, however, does genetic counseling full time.

One other point: I generally counsel in the distinctive garb of a priest. Many years ago, at Johns Hopkins University, I wore secular clothes because the people I saw there were not specifically referred to me and because Johns Hopkins University is a secular institution. Furthermore the reproductive alternatives available to Catholics at that time were perceived to be narrow indeed and the alternatives presented might be expected to observe that narrowness. Georgetown, on the other hand, is a Catholic institution, and referrals are made to me by physicians (or others) who are aware of my religious status. Also the clerical garb is an advantage. Those who come for counseling, be they Moslem, Jew, or Dunkard Brethern, expect me, as a clergyman, to be interested in them as people and not just as clients. Catholics sometimes have greater difficulty in accepting me, depending on their relationships with priests in the past. I often see on their faces the "Oh my God, it's a priest" kind of look, an attitude which has to be worked through. But those who would naturally call me "Father" slowly shift to "Doctor."

Counseling for Specific Diseases

Background Information

In general, a patient will call for an appointment after referral by a physician. The secretary ascertains the reason for the referral and the reproductive history of the person calling, if that is pertinent. The necessary information is recorded, along with an idea of when an appointment could most conveniently be made. I then review the request, match the request with records if available, decide whether to speak directly with the client or with the physician, and then assign a date, time, and counselor.

Essential to all other parts of counseling is the diagnosis. Often we request a more accurate or more specific diagnosis before the interview is scheduled. For example, we recently delayed initiating counseling for a brother and sister, both of whom were reported to have myotonic dystrophy. We wanted to get the report on the parents and determine whether one of them also had myotonic dystrophy in order to be more sure of the siblings' diagnosis. (Both the sister and the brother's wife were pregnant.) It is not uncommon to receive referrals from a very astute obstetrician who questioned the vague diagnosis presented by the patient and was wise enough to suspect some deeper problem. Thus, we recently saw a woman referred to us because of problems of infertility. She brought with her a diagnosis of mild cerebral palsy, a diagnosis that she had carried for many years. She had been operated on for orthopedic problems several times and she presented a physical appearance somewhat divergent from the norm. In taking her history, we discovered that she had a sister with Pierre Robin syndrome. In further history taking she described a number of cousins, uncles, and aunts as "uncoordinated." We referred her to a neurologist, and after an extensive workup the diagnosis came back as hereditary spastic paraplegia. This rare, dominantly inherited condition was really what we had to address in counseling.

It is not uncommon for physicians to miss a diagnosis for very rare conditions. They often treat illnesses symptomatically, and a person with a relatively rare condition almost always has received excellent symptomatic care. The point: Accurate genetic counseling can only follow accurate diagnosis. Therefore, the genetic counselor must be appropriately trained to recognize or suspect an incomplete diagnosis and must use appropriate medical referrals to specialists to get an accurate diagnosis.

The Interview

After diagnostic procedures, the client or couple is then scheduled for counseling. Genetic counseling most often involves a couple, but I also see affected individuals for counseling, often with their prospective spouses. I see healthy children in a family with one affected member or, again, a couple where one or both members have a positive family history of congenital defect. The interview begins with a cross-check on the diagnosis and the various symptoms recognized by the client. We encourage clients to talk about their perception of the problem. Thus the counselor is able to perceive the client's understanding of the condition. Generally, we see couples, and most often the woman shows the greatest immediate recognizable concern. Consequently, she is chosen as the one from whom we begin taking a medical history and pedigree information. We choose the woman also because we have observed that men are often not as familiar with their family medical histories. Accordingly, a little time is given to the men so that they can reflect on any pertinent information.

The pedigree has scientific merit in that it tells the counselor something about the patterning of disease in a family, as well as severity, age at onset, and many other helpful bits of information. Often in a classic disease, such as cystic fibrosis, little of scientific interest can be expected. I still take pedigree history, however, to rule out a second or third condition present that is not recognized as genetic. For instance, patients who have had a child with spina bifida tend to tell the counselor everything about spinal abnormality of any type and hydrocephaly, but they tend to ignore other conditions in their family histories because of their focus on this one particular problem. To encourage a more complete disclosure I use an example from many years ago. A couple referred to us had a girl who was diagnosed as having Turner syndrome, i.e., having only a single X chromosome (XO). Such a girl can be predicted to have little ovarian function and to be short. After a pedigree history and further laboratory studies, the family was told that the risk of recurrence of Turner syndrome was extremely low. The following year the father called me and said that they just had a boy and he had hemophilia. Going back to their pedigree, we noted that two cousins of the mother had died in their teens after playing sports. Both deaths sounded like sports injuries; nobody had mentioned hemorrhage; nobody had indicated hemophilia. It was a family secret kept until this boy was born. Subsequently, the girl with Turner syndrome also was diagnosed as having hemophilia. I use this example to encourage couples and counselors to be open and thorough in genetic counseling. Even more important, the pedigree is informative about the individual and couple. The counselor must find out who is feeling guilt and how each person regards his or her family. How does the couple interact? Is there support or rivalry between husband and wife?

I remember seeing a woman who had a son with an X-linked condition. She spent about 15 minutes telling me about her husband's side of the family and the many problems that had occurred on his side. She was obviously telling me how much she was suffering because her son had inherited this X-linked disorder. One also observes one member of the couple scoring points against the other, such as, "Didn't your sister have a miscarriage?"

Even if stated with a little smile, such a comment says, "That's something against your side as compared to my side." Reactions, inflections of voice, and body signals give the counselor an idea where the couple is in their marriage and, especially, who brought whom to the counseling interview. This exercise is of great importance as an indicator of the way the remainder of the conference will be shaped. Since a family pedigree is such an intimate thing, it is also the time when the sensitive counselor establishes close rapport and probes only those things of importance and reacts properly to those situations where there is humor, sadness, or shame.

After the pedigree is taken, the condition that the clients have come to learn about is explained. Most people who come for counseling think that the genetic disease entity is only that array of symptoms and effects that they have experienced or seen in a close relative. Ordinarily, any particular genetic disease has a wide range of expression, and each client usually has very limited experience with such a disease. Spina bifida, for example, is a congenital condition that can express itself either as anencephaly (no skull and only a portion of brain) or as a very small skin-covered lesion at the base of the spinal cord with an arrested hydrocephaly. The client must understand the full range of the disorder and not be overly pessimistic or optimistic. Next the cause of the condition must be explained, e.g., as lack of an enzyme, change in a structural protein, or interference with the normal process of embryogenesis; thus the patient receives a realistic view of the condition. At this point in particular, many of the questions that patients have hesitated to ask their physician seem to come out, questions that may have seemed stupid or insulting to the physician's expertise. Often patients appreciate the time their physician takes to explain everything to them and hesitate to ask for a second or third explanation.

Next comes an explanation of the genetics of the condition and the risk of recurrence. The details given in this segment depend somewhat on the sophistication of those being counseled. It is essential that the statistic presented as a risk is understood. To my statement, "The risk is 1 in 4," I have had clients respond, "Oh, since we have one child who is affected, the next three should be all right." Most of us are poor risk takers, and the counselor must ensure that these matters are correctly understood. There are always two sides to any risk figure, however. The counselor can manipulate the session at this point by presenting or emphasizing one or the other side. The risk of recurrence of spina bifida, if one child has been born affected, is about 4 percent, but the chance of having a child without spina bifida is 96 percent. The client(s) must also understand that the risk of 4 percent must be added to the normal risk (about 3 percent) that everyone has of having a child with a serious abnormality.

To this point, fair, unbiased genetic counseling would be considered to be morally acceptable to anyone. At this juncture, however, every genetic counselor must give information about all the medically and legally available options by which a person or a couple can avoid the risk of producing an affected child. Some of these options are self-evident, such as whether to marry or, if married, whether to have children. Other medical options include artificial insemination by donor in the cases of recessive disease or of a male-carried dominant trait. The option of prenatal diagnosis with possible selective abortion is available in a growing number of conditions. In too few cases is prenatal diagnosis with treatment possible, although now there are a growing number of treatable conditions too. Prenatal treatment will eventually become much more effective (see Chapters 1 and 2). I believe that the counselor must present these options without moral prejudice and without stating his or her opinion unless it is specifically requested. Just as I would argue that no genetic counselor should urge or recommend a normal pregnancy through means that I would consider immoral, so I cannot maintain such a position and at the same time try to persuade my client to *my* point of view. The best a counselor can be is unbiased in the presentation of genetic information, even though most would question the possibility of a totally unbiased presentation.

Moral Questions

There is a great good, I would like to think, of someone like myself counseling in a Catholic facility. Many counselees at this point bring in the moral dimension and ask a specific quesion, e.g., "What does the Church say?" "What should I do?" The difficulty with the open approach to counseling as previously mentioned is, of course, that some may be naive enough to think that because abortion is mentioned, it is therefore condoned. This conclusion is unwarranted but one that some clients might take. Many, perhaps most, couples have moral concerns about their reproductive choices.

In the secular society, genetic counseling stops at the point when the diagnosis is made and the necessary information has been presented. The client is expected to make a decision based on the counselor's best presentation and interpretation of the facts. In the Catholic facility, however, the moral concerns about reproductive choice will be voiced if the proper atmosphere exists. Pastoral experience and good judgment are essential in encouraging expression of the moral concerns, for, once voiced, they can be addressed.

I always go somewhat further in pressing the need for the couple to make a mutual decision. I review some of the things that must be taken into account in arriving at such a decision. A risk figure looks large or small, depending on one's experience. In my experience, couples with an affected child are far less optimistic on the average than those who see the risk only in hypothetical terms. What values one has and the way one looks at the concept of family are also important. For example, those who see children primarily as self-fulfillment have different reactions from those who see childbearing and rearing as a gift and a vocation. Each member of a couple has a different concern and a different need, and each partner must be sensitive to the other.

I urge couples to ask the question, "What would happen to her, what would happen to him, what would happen to us in our marriage if the poorest outcome occurred?" In the light of this theoretical consequence, the risk can then be weighed more realistically. I have seen couples with a 50 percent risk of having a mentally retarded child decide to attempt another pregnancy because they desperately wanted children. The threat of retardation did not hold the horror for them that it seems to hold for many other families. On the other hand, I have seen couples at a relatively low risk for minor problems decide not to have a child. Only the couple can weigh the personal significance of the facts involved because the important facts are more subjective than objective.

I never fail to emphasize to the couple that the decision they make must be mutual. Everything I have learned about the couple comes into play at this point. The dominant member must realize that mere acquiescence by the other is not adequate. The decision to have or not to have children will change their lives both as a couple and as individuals. Loving acquiescence is a way of getting along in most marriages, and as long as each partner gives in to the wishes of the other with some equity, things seem to work out reasonably well. A decision on children, however, is far too important to give in to, no matter how lovingly. A special child needs both parents, and if one has merely acceded to the other's wishes in producing this child, then that partner will not easily accept the responsibility of special care. Both must decide and both must be responsible, even if the condition is traceable to one parent.

It has been my constant belief, supported now by years of experience, that couples who properly approach this decision have a better-than-average marriage. They have been forced to search within themselves and each other and to make a very difficult mutual decision. The insensitivity of one or the other member often expresses itself at this point and progress toward greater sensitivity can be made.

Genetic counseling should be carried out before pregnancy and even before marriage. Genetic counseling in itself and in the abstract is morally indifferent. But the decisions the couple makes after counseling are full of moral content. Counselors must recognize this moral element and, although not usually trained to address it themselves, they should refer patients to a moral counselor who is aware of the area of genetics, who understands the risk figures, and who can share the implications of the diagnosis. This moral counselor may be as directive as his/her tradition demands or as experienced pastoral counseling warrants. In a Department of Obstetrics and Gynecology heavily concerned with perinatology, we receive many requests for prenatal diagnosis and care. Here is one example.

A woman aged 29 years is about 12 weeks pregnant. She is worried to some extent, but not greatly. Her uncle, a retired physician, has urged her to seek counseling and advice because he is aware of her family history. The woman's mother's first child, whose sex was unknown to the client, was born with severe malformations and died at birth. Her mother's second pregnancy resulted in a boy, who subsequently had a perfectly normal child of his own. The third pregnancy resulted in a girl, who as an adult is only 4 foot 4 inches tall and has either achondroplasia, a dominately inherited disorder (see Chapter 1), or nanism, a type of dwarfism inherited recessively. This third child was also born with a cleft palate and several other problems which make nanism a more likely diagnosis. The mother's fourth pregnancy resulted in a stillbirth or a late miscarriage. The client was the fifth child; she was followed by a stillborn and four consecutive miscarriages. Our patient also told us about a cousin who had hydrocephaly and mentioned that her grandmother and great-grandmother were short. This history, as she related it, tells us that she, or her physician-uncle, is worried about dwarfism, hydrocephaly, and perhaps miscarriage and stillbirth. She attests that any information she receives or any diagnosis of abnormality in the baby she is carrying will not alter her intention of delivering. For her, abortion is out of the question.

For optimal genetic counseling in this case, more information is necessary, i.e., an accurate diagnosis of her sister, the sex and diagnosis of her cousin, and any information available about her siblings who did not survive. Gathering such information is often painful and an intrusion on the privacy of various people, and if no other information were forthcoming, the following could be stated.

The dwarfism described could be achondroplasia, and therefore is a dominantly inherited condition. The sister would be a new mutation, as are 80 percent of such individuals, and the risk for the client's child would thus be negligible (1:250,000). If the condition were recessive (nanism), then the risk would be about 1:600. Fetal sonography could measure head size and femur length to confirm normal development in size. Hydrocephaly also can be discovered by sonography, but early detection is often impossible. Nevertheless, on the basis of the available history the risk for this client is very low. The present lack of evidence for hydrocephaly elsewhere in the pedigree should be reassuring. Repeat sonography might be suggested if there were high risk or great concern. The history of miscarriage, stillbirth, and abnormality among the siblings suggests that one of the client's parents could be a carrier of a chromosome constitution that causes such events (translocation, perhaps inversion). Studies of her chromosome makeup processed from a peripheral blood sample would give sufficient information on this possibility. Between 30 to 50 percent of the time, miscarriages are caused by the wrong amount of genetic material to the child(ren). A person with such a chromosomal makeup might lose 50 percent or more pregnancies, while the general population has a fetal wastage rate of 15 percent.

Therefore, most of the worries caused by the family history can be allayed—unfortunately not reduced to zero—by both statistics and physical evidence. Support is derived from both sources, and without putting the fetus to any known risk, e.g., by performing an amniocentesis, reassurance can be offered. Genetic counseling is sought for a particular problem of a specific family, and of the many diagnostic tools available, there will often be something appropriate to each problem. The client's intention seldom limits the type of information the counselor can give.

Preamniocentesis Counseling and Abortion

Recently, what I think of as a new type of counseling has emerged. It concerns the increased risks of mothers over age 34 and the use of prenatal diagnosis for the diagnosis of a possibly affected fetus. This approach is in some ways, similar to the screening for a specific condition in a target population, such as for thalassemia in Italians, sickle disease in blacks, or Tay-Sachs disease in Jews. But such screening is done in parents or potential parents who then make decisions about parenting or about subsequent prenatal diagnosis. Age-related screening differs in that no test is reasonable except one performed on the fetus. Today, many women realize that the risk of having a child with a chromosome abnormality leading to physical and, in most cases, mental problems increases with maternal age. (This risk is over and above the general risk of about 3 percent for congenital abnormality faced by all newborns.) Since amniocentesis is medically available and considered a reasonably safe and accurate procedure, and since diagnosis of chromosome abnormality is clinically accurate (false-positive and false-negative results being negligible), it has become necessary for every obstetrician to make sure that patients age 34 and older know of the availability of this procedure. When there is a positive diagnosis of an affected child, the only alternatives presently available are acceptance with preparation and accommodation or abortion.

The decision to have amniocentesis is a difficult one for many couples. They feel many pressures. Knowing that this approach protects them legally, obstetricians often will say, "You should have amniocentesis." The patient to whom a command has not been given, who decides not to have the procedure, and who later delivers an affected child can subsequently sue, saying "I didn't understand; the procedure wasn't explained." Many obstetricians, however, do not have the time or command of the information to explain fully. The numbers of people involved are enormous, and their problem is the same. Therefore, a special area called preamniocentesis counseling has evolved. This type of counseling is done almost exclusively in conjunction with a clinic that does the amniocentesis and the subsequent laboratory study.

Who can or should be involved in preamniocentesis counseling and laboratory diagnosis? First, who are the users? There are those who come because their physican told them to; there are those—not few in number—who are informed and want the test; and there are those who do not want the test but are being pressured by their peers and by their families. Second, who should do the counseling? Those with no qualms about subsequent abortion? Those who expect a positive result to be followed by an abortion? Those who believe it criminal for anyone knowingly to carry a defective child? The answer is obvious to those who believe in the sanctity of all life, but it means that they must be involved in a procedure that may lead to abortion. Counselors simply give information, but some couples will use this information for ill. Nevertheless, I believe that the information must be given in as unbiased a manner as possible.

Preamniocentesis counseling for age-related risks is of great moment to each couple but boring for the counselor, who is acting far more as educator than counselor. It is boring because it is repetitious. Therefore, our clinic has adopted a class format to save time and wear and tear on the counselors and, at the same time, to help the clients.

Appointments are made for about 12 couples on a Friday afternoon or a Saturday morning. They are asked to send in a family history so that we can construct a pedigree before their appointment. The class has three parts, generally handled by three different people. The first part is an introduction to explain prenatal diagnosis: what it can and cannot do, what tests may be run, and what the tests will tell and not tell. The tests are rather specific, and many patients think that at delivery they will certainly have a normal child if the tests are normal. They must be told that the risks are being lowered only slightly. The second part of the session is given by an obstetrician, who may be the one to do the amniocentesis. He or she describes the process and what to expect. Third, one of the counselors discusses the laboratory analysis and the procedure and gives both the accuracy levels and the potential risks to mother and fetus. Questions are often lively, and the group is always interesting. After class each couple is seen privately to check the family history for any other recognizable genetic condition which might need further study. All personal questions are answered. At the end of the process, couples can decide if amniocentesis is a reasonable procedure for them. Their decision will be based on information received and not on command or pressure.

Couples can be divided into four groups: those who want the procedure and are not sure what they will do with the information; those who want to know but have no intention of having an abortion; those who want the information and will abort if there is a poor outcome; and finally those who decide that any risk to the fetus (or expense) is not reasonable. In our counseling procedure, the rare patient with abnormal results is disconnected from automatic abortion by our offer of further counseling if desired and by the necessity of going to some other facility if abortion is desired. Are all the abnormal children discovered by amniocentesis aborted? No, but the majority are, for a variety of reasons.

Because Georgetown is a Catholic facility, many feel free to ask questions not usually posed. The questions include, "Does God look differently on aborting a Down syndrome baby rather than a normal baby?" "What does canon law say about abortion?" For example, a woman carrying a baby with trisomy 18, which is lethal in utero or by 1 month of age in almost 100 percent of cases, and where the baby is already compromised at 19 weeks of gestation, asked, "What does the Church say I should do?"

Today, those involved in genetic counseling must be resolved to take the best possible course of action in a less than perfect environment. For those in the Catholic health care apostolate to stay away from genetic counseling and prenatal diagnosis is to condemn large numbers of people to a situation in which amniocentesis is the rule and in which abortion is the logical and exclusive follow-up course to a less than perfect report. We, as Catholic genetic counselors, can assist in preserving the autonomy of the individual by presenting other options. If we are not involved in this field, then only those who accept abortion will be involved and heard.

Discussion

Q   My question has to do with counseling and amniocentesis. I wonder if you could expand a bit on whether a person can really remain truly neutral in a counseling session. Since you are dressed as a priest and are functioning in a Catholic hospital, may not your own moral aspects be displayed under these circumstances?

A No one can be totally neutral in presentation of any information that will ultimately relate to moral decisions. The people who come to us know, in general, where Georgetown stands on this issue, and I presume, they have a fair idea of my own position. Does my moral stance have an effect? I hope so! Do I withhold information or bias information by what I say or how I say it? I hope not! Many obstetricians opposed to abortion have, in the past, farmed out to other physicians their patients who wished amniocentesis. At 16 weeks of pregnancy, they are sent to someone who will do the procedure and the abortion when requested, or, if all is well, the patient will be directed back to the "unsullied" physician. Now, by sending the patient to us, the referring obstetricians remain involved because both they and we are involved with the outcome and subsequent counseling and referral. Some will consider the process distasteful, but I believe it witnesses to Catholic beliefs more fully than does the type of abandonment previously practiced.

Q You state that genetic counseling must be morally neutral. A Catholic hospital can make the decision whether to offer genetic counseling. You seem to imply that a Catholic hospital should have a program of genetic counseling. At what point must those who come for counseling ask for moral counseling?

A The genetic counselor should give moral counseling only if the patient asks for it. In other words, one presumes that the patient has not come seeking a Catholic attitude toward what may be done with the information provided. We could do so, of course, but we would have very limited referrals. When patients ask me for a moral opinion, they very clearly say, "Take off the one hat and put on the other." I am more than happy to do so, but only on request. On the other hand, I do point out to them that they should perhaps consider seeing their minister or their rabbi before coming to their final decision. I make sure that they understand that there is a moral dimension to the decision they are making. In general, they have not come to me to ask what my moral position is on any of these issues. They came for *genetic* information. My point is that everyone should seek to be morally neutral in counseling, because if I am not neutral in the information given, then the counselor down the street can urge things that I consider immoral.

Q I have accepted the issue of neutrality, Father. But since you have been totally honest and unbiased as best you can in making a morally neutral judgment about the genetic problem first of all, then, if questioned, you may very properly suggest moral counseling; you thus take off the neutral hat and become a moral counselor. You have no problem with that?

A No, and I think what we do is to invite the inquiry indirectly. In other words, since you have built a good rapport, since patients trust you, and since they have found you knowledgeable, they often will ask deeper personal questions. We are dealing with a biased population. The persons seeking genetic counseling tend to be better educated. Because genetic counseling represents a special but expanding level of care, it is used much more by people in a relatively affluent bracket. Among the uneducated and the poor, I see a great problem because of the coercion that can readily take place in this area. These people do what the physician tells them to do, and physicians feel much freer in saying "do this" because they are much more paternalistic toward the poor. In our population, we find that the less complicated the clients' life style, the more likely they are to decide to refuse amniocentesis after counseling.

Q This question concerns the statement that because a hospital is Catholic, it does not perform abortions, but that there are alternatives. Is that information communicated within the context of a geneticist and not a moral advisor? In other words, people are coming to you as a geneticist in a Catholic institution. Are you the one that imparts that information?

A Almost always. There are people on either side of the issue who do not want to talk. I remember talking to one woman who had been trying to get pregnant for the last 6 or 7 years. This pregnancy was perhaps her only pregnancy. Amniocentesis revealed that she had a Down syndrome child. That woman would have certainly had the baby, but she was married to someone who had two sons living at home, both with a lethal muscular disease, and both in their teens and dying. She felt that the burden was just too much. In the midst of her tears she said, "I would happily have this baby except in this circumstance." We can sit down with her and offer further counseling.

We saw another couple before the wife became pregnant; they had decided to have a child even though she was 38 or 39. Subsequently, they came in for amniocentesis, and a high alpha-fetoprotein level was found, which generally indicates an abnormal opening of the fetus. Sonography showed the high level to be caused by an abdominal herniation. They accepted the diagnosis, and then the chromosome analysis came back as trisomy 18, which most of the time leads to spontaneous miscarriage. The small percentage of children who reach birth do not survive; 95 percent die within the first month. The couple was very upset, and the wife asked me what the Church would recommend. Although she was a nonpracticing Catholic (and had not practiced for a long while), I could tell her what, heroically, I thought the Church would ask her to do. I also knew what she was going to do; interestingly, after she had the abortion of a fetus who probably would have died in any event in 2 or 3 weeks, she asked whether I knew someone who would come and say a prayer. The couple wanted to bury the fetus someplace where they could visit. You can see the pastoral problem that this woman is going to have in the future. We are very aware of the kinds of problems that people face. People are sometimes very cold about what they will do, but most present pastoral problems no matter which decision they make. We offer the subsequent counseling and go with them as far as they permit us. At this point the counselor might decide what the couple is going to do no matter what counsel is given, and the counselor must help them draw out the best from their unfortunate situation. The answer to the question is that we see as many people as will come back to see us.

Q Father, I assume from your discussion that amniocentesis is performed at Georgetown University Hospital. What would the feasibility be of having that procedure performed at a non-Catholic hospital?

A The feasibility? There is no problem going to a commerical laboratory which will take as many people as you send there. But a commercial lab will not do as good a job and will do absolutely no counseling. Commercial laboratories are now set up for a number of prenatal diagnostic tests. Other university laboratories are also available in the Washington, DC, area. I believe, however, that the patient gets more information and has the opportunity to make a better informed decision with us than with any other facility locally available.

chapter 4

Genetic Screening and Prenatal Diagnosis

Robert F. Murray, Jr, MD

Rationale and Structure of Genetic Screening Programs

The underlying purpose of genetic screening is intervention. As in medicine in general, those involved in genetic screening intend to intervene in a process, usually a disease process. They hope to identify the possible illness before it upsets body equilibrium or homeostatis. In short, genetic screening is a type of preventive maintenance. The idea behind genetic screening is to devise a means of testing to find and repair potential damage before it becomes widespread, expensive, and irreversible.

Tests

The best screening programs select from a healthy population all those with potential disease, without exception. If the test is sensitive enough not to miss a person with the slightest chance of developing the illness, this usually means it will produce some false-positive results. This requires a second more definitive test to weed out the people who had false-positive results. This second test must be a more accurate test and thus is usually more expensive. Above all, false-negative results must be avoided. One of the tragedies of genetic screening is finding out that a person has a condition that the test was supposed to pick up but did not. In the best of all possible worlds there would be no false-positive or false-negative results. Such a test

---

Dr. Murray is chief, Division of Medical Genetics, Department of Pediatrics and Child Health, Howard University College of Medicine, Washington, DC.

would minimize or even eliminate all errors involved and be a perfect screening test. Unfortunately, that kind of test does not yet exist.

There are two other important ingredients in effective screening tests, whether a test is used to screen for diabetes or for a genetic disease. One, the test must be inexpensive. Two, it must be simple and require very little in the way of expensive equipment. One goal of most health care programs is saving money. Genetic screening, likewise, exists not only to avert patients' pain and suffering but to save money. Economists and administrators have devised a cost:effectiveness, or cost:benefit, ratio, which compares the cost of finding one case of the condition being screened for with the cost of allowing the disease to develop and the affected persons to live out their lives in institutions.

Charles Scriver, MD, of Montreal, a pioneer in genetic screening who has spent many years in this field devising and refining biochemical tests for inborn errors of metabolism, has an interesting way of illustrating the concept of screening. He has a slide of a poster showing a large number of men wearing blue hats, all of whom seem to be looking in one direction; somewhere in this field of men is a small group of men wearing red hats and in this group is a man wearing a red hat *and* looking in the direction opposite from everyone else. The ideal screening test would be able to find that one man in the red hat who is also looking in the opposite direction *every time without fail* and do it in such a manner as to cause minimal or no discomfort to the people in the blue hats, and those few in red hats looking in the same direction as everyone else.

If possible, geneticists would like to have a *qualitative* test, i.e., a test whose result is a simple yes or no. Unfortunately, most of the available tests are *quantitative*. This means that there is a range of normal values, the values found in the so-called healthy population of people who do not have a specific disorder or disease. Then workers define cutoff points, usually by a standard statistical definition (e.g., two or three standard deviations outside the normal range). Almost always, there are some people with the disease whose results fall in the *normal* range and some people who are *healthy* whose results fall in the *abnormal* range. Sometimes these people have other conditions that cause this discrepancy. Factors such as these help account for false-negative and false-positive results.

Another problem is identifying the target population. Genetic diseases, like most diseases, are not distributed uniformly in all populations or ethnic groups. A target population, i.e., the population where the condition appears to occur with especially high frequency, must be defined. Screening efforts that focus on the target population with the greatest risk are most likely to be cost effective.

Purposes

Genetic screening programs differ from other screening programs in at least two respects. First, in addition to the goal of early identification, such programs often focus on people who are really never going to be sick themselves but who are at risk to have children who may be sick. These people are called genetic carriers. Second, screening may focus on finding the disease in fetuses. I would like to add another point here: One of the consequences of identifying the disease at the fetal stage is that rarely, the disease which is fully developed or almost fully developed at the time the child is born, may be prevented if it is found and treated early in gestation.

The report of the National Academy of Science's Committee on Genetic Screening recognized three major purposes for genetic screening: (1) to detect affected persons to treat in order to avert the consequence of disease; (2) to detect carriers of genes or carriers of affected fetuses to provide reproductive advice; and (3) to screen for research purposes.[1] These purposes are not mutually exclusive.

One question about screening that is still unanswered and raises legal, ethical, and moral concerns is, When does a genetic or a diagnostic test or a prenatal diagnostic test cease being a research procedure, where it is being tested for its efficacy and safety and become a part of the accepted medical armamentarium?

There are several periods in human development at which genetic intervention might occur. These are (1) with the gametes, i.e., the eggs or sperm that produce the human being at fertilization, (2) at the level of the fertilized egg (or zygote), where some people feel scientists will begin to make the initial impact through genetic engineering of affected individuals, (3) at various stages of embryonic development, especially before the fetus is fully formed, (4) at 12 weeks of gestation and beyond, (5) at birth, in concert with newborn screening, (6) in childhood, and (7) throughout young adulthood, where reproductive counseling as well as medical management are important. Finally, intervention may involve influencing the act of reproduction, by helping couples to decide *whether* to have children or *with whom* to have children.

Some of the reasons for genetic screening are as follows:
- Early diagnosis of the homozygote, i.e., of the affected fetus, and subsequent treatment.
- Heterozygote detection, i.e., detection of the unaffected carrier. Genetic counseling, of course, follows the detection of carriers, especially in couples at risk.
- Prenatal genetic diagnosis and selected abortion where desired by the parents.

- Identification of older women or men who are thus at higher risk than the general population, e.g., providing amniocentesis to women over 35 years of age.

I have purposely not discussed reducing the frequency of abnormal or mutant genes through screening programs because of the unethical aspects of so-called negative eugenic programs. These are based on the idea that some genes are clearly detrimental to humans and that society should therefore make an overt effort to prevent them from being passed on to the next generation.[2] Except for those genes that produce grossly abnormal function and malformation, genes are not inherently bad. Geneticists are beginning to find, as in the case of the sickle cell gene, for example, that a gene which in a double dose produces disease may, in a single dose, protect people from serious disease caused by environmental hazards. Genes may also have other as yet unidentified functions. It is difficult to label a gene good, bad, or even normal except on a statistical basis. A gene considered abnormal in the U.S. population is considered normal in other populations, since most people in that population are carriers of that gene. Trying to define normal and abnormal or good or bad in genetic terms becomes very difficult.

Settings

The settings for screening programs may, quite appropriately, differ. The location of screening may be determined by program type and scope and necessary follow-up facilities.[3] For example, in newborn screening programs, initial samples can be taken in hospitals, but the follow-up or confirmatory specimens may have to be taken in the private physician's office. Screening specimens from genetic carriers may be taken in a variety of places, e.g., community neighborhood health centers, churches, mobile health units, family planning clinics, and physicians' offices. Screening for the presence of affected fetuses in pregnant women usually takes place in a private obstetrician's office but may take place at the family practitioner's office, family planning clinic, or special hospital clinics. Each type of program, because of its specific clientele, special focus, or special needs, is pursued in a different setting, and programs may take place under different auspices. For example, if the state were interested in certain programs, public health departments would promote screening. On the other hand, a city or even an ethnic group might be interested in other programs, as in the case of a Tay-Sachs screening program, and may determine that it is in its interest to promote and finance screening.

Selected Conditions

The prototype of screening programs in medicine has emphasized early detection of affected persons so that they can be treated. This is in keeping with the medical principles of relieving suffering and disease. The initial thrust of genetic screening was to detect children who would manifest some degree of developmental retardation or a consequence of an inborn error of metabolism. The prototype disease due to an inborn error of metabolism is phenylketonuria (PKU).[4, 5] In brief, it is a disorder in which the infant, normal at birth, remains fine until it begins to feed. Then, an amino acid called phenylalanine, a normal component of the diet, cannot be properly metabolized by these infants and builds up in their blood. It, or its metabolic by-products, cause brain damage. The brain does not properly develop, and the children become moderately or severely retarded unless they are identified within 6 weeks of birth and placed on a special diet low in phenylalanine. PKU is relatively infrequent among the important childhood diseases, but it is one for which there is an effective and inexpensive screening test as well as an effective treatment. A number of other genetically determined conditions are much more frequent in children, including sickle cell anemia, cystic fibrosis, and diabetes. Why, then, choose to screen for disease that is this infrequent. (In most populations, the condition is seen in 1:15,000 infants.) Because PKU is treatable and because the screening test for PKU is inexpensive. In addition, PKU screening can be combined with a test to detect a number of other, less common metabolic errors.

TABLE 1: U.S. Frequencies of Some Metabolic Disorders and Birth Defects for which Screening Tests are Available.

Disorder or Defect	Frequency
PKU	1:15,000 U.S.
Maple syrup urine disease	1:200,000
Homocystinuria	1:220,000
Galactosemia	1:75,000
Congenital hypothyroidism	1:4000-1:8000
Tay-Sachs disease (Jews)	1:3600
Sickle cell anemia (Blacks)	1:600
Cooley's anemia (thalassemia major) (Mediterranean)	1:1000-1:2000
Neural tube defects (Europeans)	1:500-1:1000

More frequent genetic disorders in particular populations are shown in Table 1. Tay-Sachs disease[6] occurs in 1:3,000 to 3,600 Jews of eastern European origin at birth; sickle cell disease, in 1:600 or 700 blacks; cystic fibrosis, in about 1:1,600 whites. Targeting the right population for screening becomes important in a cost-effective screening program; however, if one can identify affected children, treat the disease, and produce a useful, healthy citizen, the effectiveness of that program is magnified even more, since it has added many years of useful life and productivity to that person. Even though these specific populations have these diseases with this frequency, however, it does not mean that other groups do not have those diseases at all. One of the questions to be addressed in screening programs is what to do about people who do not fall in specific high-risk groups. Sickle cell anemia does occur in whites, although uncommonly. Tay-Sachs disease can occur in non-Jews. In fact, in some communities it is now being seen more frequently in non-Jewish infants, partly as a result of the success of screening programs set up to prevent its occurrence in Jews. People other than Malaysians and Chinese have alpha-thalassemia, a severe, sometimes lethal blood disease. And the list goes on. Even though it is not cost effective to involve people who are not in high risk groups in screening programs, do they not have the right to access to screening programs?

At least seven screening tests for metabolic diseases can be done from a sample of blood taken either at birth from the umbilical cord or on the second or third day after birth from a heel prick. Most of these are tests for conditions that can be diagnosed from a drop of blood, which is placed on a piece of filter paper which, when dried, can be stored until it is mailed to a central processing laboratory where these tests can be carried out. Massachusetts has been carrying out such screening tests for various metabolic conditions for almost 20 years.[7,8] Many of these disorders can be treated, some more effectively than others, to avoid mental retardation and early death. In some cases (e.g., galactosemia), infants are susceptible to infection very early. Usually, by the time the physician is aware that they have a certain metabolic condition, they have developed septicemia and meningitis and may die. Early detection here can save a life; in this case, the avoidance of galactose in the diet is very effective in preventing the disease and its complications. There are many other potentially detectable conditions, however, for which no effective treatment or medical management is currently available.

Another genetically related disease suitable for newborn screening is congenital hypothyroidism.[9] This disease is caused by an absence of the thyroid gland, an abnormality in the pituitary gland, or an inherited genetic abnormality in thyroid hormonal production. It occurs in approximately 1:4,000 to 1:8,000 newborns, depending on the population tested. This condition is very simple and inexpensive to treat. One merely gives the infant thyroid hormone by mouth. If treated properly, infants develop normally and retardation is averted. Approximately 39 states now require or strongly suggest testing for PKU and for hypothyroidism. In Washington, DC, a recently passed law has made testing for PKU and hypothyroidism mandatory for all infants born in hospitals. Parents can refuse the test on religious or ethical grounds if they wish, but they must sign a statement in order for their baby *not* to be tested.

Starting and operating an urban newborn screening program is a difficult task, however. One major job is getting all the hospitals involved to collect specimens. One practical consideration is deciding who pays for collecting the samples, because the city generally only pays for the tests that are performed. Moreover, what about repeat or follow-up tests? In the case of PKU, 95 percent of the first positive tests are false-positive results because at the time when the baby must be tested—the first 2 to 3 days—the cutoff point for abnormality must be set very low. A second test must be carried out to distinguish affected infants from those with false-positive results on initial screening. Who pays for the important second test? Many parents cannot afford to pay for the test. Some parents have a private doctor or do not get back to the clinic for that second test. This is another vital question to be considered in a mandatory testing program, particularly a large-scale program.

In the case of sickle cell anemia, for example, there is no definitive treatment. The basis for screening in the newborn period in sickle cell anemia is to provide better medical management designed to prevent complications from infections and sickle cell crises. Based on current knowledge of the natural history of this disease, many children with sickle cell anemia may die by 2 or 3 years of age because their physician does not know they have the disease. By the time the definitive diagnosis is made, the child may be critically ill.

Screening for Reproductive Advice

Tay-Sachs Disease

The first model of a screening program for reproductive advice that I will discuss here involves prenatal detection of Tay-Sachs disease.[10] Tay-Sachs disease, one of a group of metabolic conditions called the gangliosidoses, is caused by a deficiency of an enzyme called hexoaminidase A. It is much more frequent in Eastern European Jews or their descendants than in other ethnic groups. Even though it does occur in other ethnic groups, it occurs in 1:3,000 to 1:3,600 Ashkenazic Jews but in only 1:36,000 of the general population.

Michael Kaback, MD, spent many years and built his outstanding career on developing a prototype screening program for the carriers of the Tay-Sachs gene.[11] He did this by working intensively with the Jewish community. In most other screening programs, screening was initiated by the public health department or with private physicians; Dr. Kaback, however, went directly to the Jewish community. He educated them about (1) the nature and the frequency of the disease, (2) the fact that there was a prenatal diagnostic test available to diagnose the condition, and (3) the fact that 70 to 75 percent of couples with an affected child would not risk having another child. Children with Tay-Sachs disease appear healthy and live normally for 3 to 6 months, after which they begin to deteriorate to a vegetative condition. By 3 years of age they are totally incommunicative, develop malnutrition and/or infection, and they die. With the prenatal diagnostic tests, however, parents who can morally accept abortion have the option of terminating this pregnancy.

Dr. Kaback recently reported on his worldwide experience in Tay-Sachs testing.[12] About 250,000 carrier tests have been carried out; 210,000 have taken place in the United States. About 10,500 carriers have been detected worldwide, with about 8,000 of these in the United States. These figures were collected from 1969 to 1979, and out of this total screening experience, 210 couples at risk were identified. First the wife is tested; if she is a carrier, her husband is also tested. If both are carriers, and if the couple wishes, their pregnancy is "monitored," i.e., fetal cells are tested for the disease. Of the 210 couples at risk, roughly 180 pregnancies were followed. About 32 infants were diagnosed to have the disease and 31 pregnancies were terminated. Among 450 couples who already had at least one Tay-Sachs child, 116 had affected fetuses; 111 of those parents elected to terminate the pregnancies. In counseling these couples, Dr. Kaback emphasized that the use of prenatal diagnosis does not imply a couple's commitment to abortion if an affected fetus is identified. Fr. Baumiller (see Chapter 3) noted that some places put pressure on couples to commit themselves to abortion if a test is positive. Dr. Kaback does not. There are very strong Orthodox groups in the Jewish community who are totally opposed to abortion under any circumstances. This program could not have started if abortion was the premise on which it was based. Dr. Kaback made it clear that couples should have the option of prenatal diagnosis without first making the decision to terminate; 5 of 116 couples with affected children decided not to terminate their present pregnancies.

The "true" efficacy of this program is being questioned at this time.[13] If one can, for the moment, put aside the moral question of abortion, one might say this is an ideal preventive genetic program. A very good test is being applied to a defined target population. There is a definite action that the couples can take to avoid having affected children. The evidence from Dr. Kaback's sociological studies suggests that couples at risk who would not have had children without the availability of the program decided to have children. He concludes that this program has been positive rather than negative.

Sickle Cell Anemia

Sickle cell anemia occurs in the United States in roughly 1:600-700 black neonates.[14] The sickle cell trait, in which the person carries one sickle gene but is otherwise healthy, is seen in about 1:12 blacks in the United States. These frequencies vary somewhat throughout the United States depending on intermixture with the white or American Indian population. Thalassemia major, or Cooley's anemia, another blood disease, sometimes occurs in genetic combination with sickle cell anemia. It is seen in 1:1,000 to 1:2,000 persons of Greek, Italian, or Middle Eastern ancestry; about 1:25 people will carry the gene for thalassemia.[15] When the two genes combine, they produce a condition called sickle-thalassemia disease. Both these conditions, sickle cell anemia and Cooley's anemia, occur much more frequently than other conditions currently being screened for in the population. Screening programs for these conditions in target populations could be relatively easy and effective. Unfortunately, there are no effective treatments for either disorder.

Recently, researchers have found ways to relieve the complications of chronic blood transfusion, the main method of treating Cooley's anemia, which has contributed to marked improvement in the life expectancy of these patients. Nevertheless, parents who already have had a child with Cooley's anemia are much more likely to not want any more children than parents who have a child with sickle cell anemia. These two conditions are comparable genetically and hematologically, although they are not comparable in a clinical sense, both are rather frequent in the United States.

TABLE 2: Incidence of Sickle Cell Diseases in the U.S. Black Population

Condition	Hemoglobin Type	%
Sickle cell anemia	SS	0.3-1.3
Sickle cell hemoglobin C disease	SC	0.1-0.25
Sickle cell thalassemia disease	S Thal	0.04
Sickle cell hemoglobin D disease	SD	0.0083

Table 2 lists different kinds of sickle cell disease and their frequencies in the U.S. Sickle cell anemia is clearly most frequent. Combinations of sickle genes with other abnormal hemoglobin genes, however, produce sickle cell diseases that, although clinically less severe than sickle cell anemia, are no less troublesome from the standpoint of the parents. Some of these disorders can be very mild, and patients' life expectancy in some cases is within normal limits. Afflicted persons may have some clinical difficulty but they can also live to be 60 or 70. There are a few reported cases of people with sickle cell anemia who live to be 70 years of age.

The first screening test for sickle cell disease, the one that put sickle cell screening on the map, was very simple, almost too simple, precipitation test. A special solution of dithionate is put in a test tube at acid pH. A drop of blood is added, the tube is shaken, allowed to stand for a few minutes, and is read against a specially lit, lined background. When the solution turns cloudy and the lines cannot be seen through the tube, the test is positive; when they can be seen, the test is negative. The test requires no special equipment and no special training is needed. The testing solution is very inexpensive. The problem with the test, however, is that a positive result does not distinguish between persons who have sickle cell anemia and persons who carry the sickle cell gene. There was as a result a great deal of panic in the black community for 2 or 3 years until the more acceptable and definitive electrophoresis test was introduced. This test uses a small blood sample on special paper and a power pack that produces electric current that flows through this paper. This procedure separates different hemoglobin types, making it possible to distinguish between persons who have the disease and those who are carriers. Even this test is not perfect, however, since 30 or so different hemoglobin types have the same mobility as sickle cell hemoglobin. Most of these different hemoglobins can be distinguished by using a combination of the electrophoresis test and the solubility test described above. If the solubility test is positive and the electrophoresis shows a single band of hemoglobin in the position for sickle hemoglobin or two bands in the positions of sickle and adult hemoglobin, the individual has sickle cell anemia or is a sickle cell gene carrier respectively.

The person with sickle cell anemia has parents who are both carriers of the gene. Two parents who are sickle cell gene carriers have a 1:4 chance of having a child with sickle cell anemia with each pregnancy. Thus, they might benefit from genetic counseling. The counselor can explain the genetics of the sickle cell trait and disease, help couples understand the difference between having the disease and being a carrier, and inform them that there are combinations of other abnormal hemoglobins, many of which, in combination with the sickle gene, can also produce a sickle cell disease. The simple screening process described above thus gets increasingly complicated. Not only must the test be able to detect the sickle gene, but rarer and more complex hemoglobin variants must be identified.

There is now a way to detect this disorder prenatally, either by taking a blood sample from the placenta and studying the hemoglobin or by studying the effect of specific restriction enzymes on the deoxyribonucleic (DNA) acid of the fetus in cells taken by amniocentesis.[16] These procedures will probably become generally available in a few years.

Other Topics of Concern

I would like to mention two other areas of reproductive counseling. The first concerns screening in pregnancy for chromosome abnormalities of the type that Fr. Baumiller mentioned (see Chapter 3), i.e., the screening of mothers over age 35, who are at higher risk to have children with Down syndrome.[17] This screening program provides information to women aged 35 and over about the increased risk of their having a child with Down syndrome. They have the option of prenatal diagnosis by amniocentesis and selective abortion of affected fetuses. There are now *legal* reasons why a mother over 35 must be told about this option. Obstetricians who neglected to tell a woman 35 or older who had a Down syndrome child about this risk and the prenatal diagnostic option have been sued successfully for malpractice. Prenatal diagnosis for women over age 35 is now becoming more important in screening. There are now several large-scale city-based screening programs for older women. Men aged 45 years or older also contribute to the Down syndrome risk in a 1:5 ratio. For instance, of every five cases of children with age-related Down syndrome, four will have been caused by an older mother and one will have been caused by an older father. Fathers can no longer be considered exempt from contributing to the nondisjunction responsible for the extra chromosome no. 21 that causes Down syndrome.

A large percentage of Down syndrome children are born to mothers under 35 years of age. An important question that should be addressed is, Do not those women who are in the lower risk group have a right to this diagnostic test if they wish? At 35 years the age-related risk is about 1:350, while at 20 to 24 years it is 1:1,550. The risk between age 30 to 34 is 1:700, which is only one-half the risk at age 35. The reason for choosing the cutoff point at age 35 is the cost effectiveness of the program. If women *under* age 35 are included, the program of screening for chromosome abnormalities by amniocentesis is no longer cost effective. This conclusion was arrived at by balancing the cost of screening against the cost of not screening.

Finally, I would like to consider programs of screening for increased levels of maternal serum alpha-fetoprotein, which is a means of determining whether a developing fetus might have a neural tube defect.[18] This is the newest large-scale reproductive screening program that has been adopted. In Scotland and Wales, 40 to 50 percent of all pregnant women are screened for maternal serum alpha-fetoprotein in order to detect neural tube defects prenatally. This fetal-specific protein produced by the liver is always present to some degree in the amniotic fluid. When it is present in high concentration, it diffuses into the mother's circulation and can be detected in her blood. A screening program to test the mother's blood for the presence of increased amounts of this protein as an indicator of a neural tube defect has been adopted in several countries, and pilot programs are running in various U.S. locations.

A major problem of this testing program is that it gives a high frequency of false-positive results. In this country, only 1 or 2:1,000 women will actually have a baby with a neural tube defect; however, when the sera of 1,000 pregnant mothers are tested, 50 of those women will have a positive test on the first trial. In a second test, 30 of those 50 mothers will still show a positive result, which is still 15 times more frequent than the true positive results. To define the affected fetus further, an ultrasound analysis can identify or outline the defective head of the anencephalic infant or spot one with a serious spina bifida anomaly, among other things.

Ultrasound can show that a woman is pregnant with twins, which increases the level of alpha-fetoprotein, that the fetus is dead, or that the estimated gestational age is wrong. This is important because the amount of alpha-fetoprotein varies with the age of the fetus. It is very high early and then declines. If the woman thinks her pregnancy is further along than it actually is, she might appear to have an elevated level of alpha-fetoprotein when, in fact, her pregnancy is not as far along as was thought. In short, an elevated alpha-fetoprotein level does not always mean a neural tube defect.

Of the 1,000 women tested fifteen women have elevated fetoprotein levels that cannot be explained by ultrasound, and an amniocentesis must be performed and the fluid tested. Theoretically, if the amniotic fluid is tested and the tissue acetylcholinesterase (an enzyme produced primarily by nervous tissue) measured, the combination of these findings should unambiguously identify women who carry an anencephalic child or a child with severe spina bifida and meningomyelocele.

Unfortunately, this procedure is not perfect. There are still some rare errors, including some where positive results might lead a woman to decide to terminate a pregnancy and the fetus turns out not to have a neural tube defect. Even though this is rare, whenever it happens it is a tragedy. Despite these rare mistakes, this screening program for neural tube defects may be the one with the widest effect in the future because the initial blood sample is so easy to take during the routine obstetrical evaluation.

The Food and Drug Administration (FDA) has just gone through an extensive program attempting to analyze the feasibility of introducing a commercially available testing kit for alpha-fetoprotein. Anyone who is qualified to buy and perform the test can have the screening program carried out in their own jurisdiction. If this screening program were introduced nationwide, it might cost anywhere from $20 to $90 million a year to administer, assuming all pregnant women participated. Some people are talking about implementing this program nationwide. By some estimates, at least twice as much in medical costs might be saved with universal large-scale screenings and if all pregnancies of affected infants were terminated.

Despite the problems involved in screening, I believe the time will come when every infant born will be subjected to a battery of screening tests to identify any disease that will require future medical care and, particularly, to allow early *treatment* in order to avert any harmful effects.

Discussion

Q You implied that maybe in some situations screening is obligatory, and you mentioned Washington, DC. It is obligatory for the newborn to be tested for PKU as well as for hypothyroidism. My question is, Do you foresee the possibility for screening to be a requisite for marriage? Most if not all states require serological testing before a marriage license can be obtained. Would it be possible to require testing for a specified genetic condition before a couple could get a marriage license?

A This is an interesting question, and one that has been broached in other states. New Jersey, for a brief period, had a program in which genetic screening was required for a marriage license, but the requirement was challenged and removed. An association of lawyers in Illinois wanted to introduce legislation that would make such genetic testing mandatory. If a couple were both found to be carriers of the same recessive gene that could produce disease, they would be denied a marriage license. This denial would probably have no effect, and most likely the couple would go to another state to get their license. Once a couple has decided to get married, genetic counseling is not likely to change their minds. In our own programs, we have chosen to emphasize screening and counseling in the adolescent and young adult period. With the increasing problem of adolescent pregnancy, genetic screening must be emphasized early.

Those states that had such laws now provide information to prospective couples who request it. When the routine premarital blood test is given, the couple is informed that genetic tests are available if they wish. If they do not refuse the tests, and if they both have positive results for a given genetic disorder, the state refers them to (and pays for) appropriate counseling facilities. I think this is an appropriate way to provide counseling, although I think the information is given too late to influence most marriages.

One study carried out in Greece attempted to influence the mating pattern in a defined population on a genetic basis.[19] In Orchomenos, a village outside Athens, extensive genetic screening for sickle cell anemia and thalassemia was carried out, followed by a very intense counseling program. The investigators carried on a *directive* counseling program, something which neither Fr. Baumiller nor I agree with, i.e., those who were counseled were told not to marry someone who also had the sickle cell or thalassemia gene. Seven years later the village was surveyed, and the investigators counted up the number of affected children born during that period. They found the frequency of both sickle cell anemia and thalassemia had not changed one iota. They then checked the mating pattern, thinking that perhaps the effect of counseling had not had a chance to diffuse into the reproduction pattern. They found that the mating patterns were essentially the same and that the frequencies of mating between people who both had sickle cell trait or thalassemia trait were the same. What the investigators had influenced was the status of women who were carriers of those genes. Women who carried the sickle cell or the thalassemia gene were stigmatized and were considered undesirable mates, not just by men who also carried the gene, but by men who did not but who recognized from the genetic patterns that their children might also be carriers. The net effect of the attempt to alter mating patterns through genetic screening and counseling was therefore negative. It indicated that trying to change longstanding cultural patterns can be disastrous.

Q You indicated in Dr. Kaback's study that of 454 people, 160 had Tay-Sachs disease and 111 opted to have an abortion. Are there any similar statistics with sickle cell anemia?

A Maybe I was not clear when I gave those statistics. One hundred sixteen women were identified as carrying an affected fetus. Of these women, 111 opted to abort, leaving 5 couples who decided not to terminate the pregnancy of a fetus with Tay-Sachs disease. There are preliminary figures for sickle cell anemia from one study that show that roughly half the couples who are informed about prenatal diagnosis decide to take advantage of it. I believe about 90 percent of couples decide to take advantage of prenatal diagnosis for Tay-Sachs disease at the first step. Of the 50 percent of people who take advantage of sickle cell anemia screening programs, half of those with an affected child, go on to terminate the pregnancy. The reason for this seems clear. Sickle cell anemia does not produce a retarded child. It results in a human being who has the potential for living a full and productive life and contributing to society. No deformity is present, and the only thing that distinguishes these people from the rest of society is that on the average they get sick more frequently and they suffer more often from severe pain. Parents who already have affected children find it more difficult to consider prenatal diagnosis and abortion than parents who do not. Physicians have not really studied the emotions and moral feelings involved on a couple-by-couple basis, but the type and severity of a genetic disorder obviously influences the couple's decision whether to continue or to terminate the pregnancy.

About 85 to 90 percent of couples who go through prenatal diagnosis for Down syndrome and find they have an affected child terminate the pregnancy. About 10 to 15 percent decide, for one reason or another, even when they already have an affected child, not to terminate the pregnancy. These figures come from clinics where there is no pressure, at least according to their public statements, to force couples to terminate the pregnancies of affected children.

Footnotes

1. *Genetic Screening: Programs, Principles and Research.* Committee for the Study of Inborn Errors of Metabolism, (Washington, DC.: National Academy of Sciences 1975).
2. A. Capron, et al., eds., *Genetic Counseling: Facts, Values and Norms.* The National Foundation-March of Dimes Birth Defects: Original Article Series. (New York: Alan R. Liss, 1979).
3. Daniel Bergsma, ed. *Ethical, Social and Legal Dimensions of Screening for Human Genetic Diseases.* The National Foundation-March of Dimes Birth Defects: Original Article Series. (New York: Alan R. Liss, 1974).
4. Neil A. Holtzman, *Newborn Screening for Genetic-Metabolic Diseases, Progress, Principles and Recommendations.* (Washington, DC: U.S. Department of Health, Education, and Welfare, pub. no. [HSA] 78-5207, 1977).
5. A. Y. Tourian and J. B. Sidbury, "Phenylketonuria," in *The Metabolic Basis of Inherited Disease,* J. B. Stanbury, J. B. Wyngaarden, and D. S. Fredrickson, eds. (New York: McGraw-Hill Book Co., 1978), Chapter 11, pp. 240-255.
6. M. M. Kaback, "Heterozygote Screening and Prenatal Diagnosis," in *Tay-Sachs Disease: A Worldwide Update in Lysosomes and Lysosomal Storage Diseases,* J. W. Callahan and J. A. Lowden, eds., (New York: Raven Press, 1981), pp. 331-342.
7. H. L. Levy, "Newborn Metabolic Screening: Past and Prospects," *New England Journal of Medicine* 293 (1975): 824-830.
8. "Massachusetts Department of Public Health: Cost-benefit analysis of newborn screening for metabolic disorders." *New England Journal of Medicine* 291 (1974): 1414-1418.

9. G. N. Burrow and J. H. Dussault, eds., *Neonatal Thyroid Screening*. (New York: Raven Press, 1980).
10. M. M. Kaback, T. J. Nathan, and S. Greenwald, "Tay-Sachs Disease: Heterozygote Screening and Prenatal Diagnosis — U. S. Experience and World Perspective," in *Tay-Sachs Disease: Screening and Prevention*, M. M. Kaback, ed. (New York: Alan R. Liss, 1977), pp. 13-36.
11. Ibid.
12. Kaback, *Tay-Sachs Disease: A Worldwide Update in Lysosomes and Lysosomal Storage Diseases*, pp. 331-342.
13. M. W. Steele, "Lessons from the American Tay-Sachs Programme," *Lancet* 1 (1980): 914.
14. R. M. Winslow, and W. F. Anderson, "The Hemoglobinopathies," in *The Metabolic Basis of Inherited Disease*. J. B. Stanbury, J. B. Wyngaarden, and D. S. Fredrickson, eds. (New York: McGraw-Hill Book Co., 1978) Chapter 62, pp. 1465-1507.
15. D. J. Weatherall, "The Thalassemias," in *The Metabolic Basis of Inherited Disease*, J. B. Stanbury, J. B. Wyngaarden, and D. S. Fredrickson, eds. (New York: McGraw-Hill Book Company, 1978) Chapter 63, pp. 1508-1525.
16. Blanche Alter, "Prenatal Diagnosis of Hemoglobinopathies: A Status Report," *Lancet* 2 (1981): 1152-1154.
17. *Antenatal Diagnosis: Report of a Consensus Development Conference*. (Washington, DC: U.S. Department of Health, Education, and Welfare, Public Health Service, National Institutes of Health, 1979), pp. 33-35.
18. B. Gastel, et al., eds., *Maternal Serum Alpha-Fetoprotein: Issues in the Prenatal Screening and Diagnosis in Neural Tube Defects*. Conference Proceedings, National Center for Health Care Technology, 1980.
19. G. Stamatoyannopoulos, "Problems of Screening and Counseling in the Hemoglobinopathies," in *Birth Defects*. A. G. Motulsky and W. Lenz, eds. Proceedings of the Fourth International Conference, (Amsterdam: Excerpta Medica, 1974), pp. 268-276.

PART II

GENETIC ENGINEERING

chapter 5

Treatment Of Genetic Diseases:
Inborn Errors of Metabolism

Seymour Packman, MD

The preceding four chapters have already addressed treatment of genetic disease, if we consider that any form of medical intervention constitutes a form of therapy. In medical genetics, a major therapeutic activity is the provision of information, i.e., genetic counseling (see Chapter 3). Efforts to understand genetic disease (see Chapter 1), to diagnose genetic illness prenatally (see Chapter 2), and to screen for genetic diseases presymptomatically (see Chapter 4) improve on the information physicians and other health care professionals give to patients and allow them to refine such genetic counseling. Treatment of genetic diseases is therefore not a new activity in medical genetics, and the topics discussed in this chapter should be viewed as part of the continuum illustrated in the list below:

- Postsymptomatic therapy;
- Presymptomatic therapy;
- Prenatal therapy;
- Prenatal diagnosis; and
- Genetic counseling.

The activities listed above are in a sequence of physicians' increasing ability to influence the course of events in genetic diseases. In the worst instance—postsymptomatic therapy—physicians are presented with a patient who already has a disease, and the task is to ameliorate the symptoms and prevent further (neurological and other) damage. Postsymptomatic

---

Dr. Packman is in the Department of Pediatrics, Division of Genetics, School of Medicine, University of California, San Francisco, CA.

therapy is a most frequent situation and gives physicians the least opportunity to influence the patient's final outcome favorably.

In marked contrast, physicians can influence the course of events before conception by informing couples of diagnoses, prognosis, and recurrence risks in genetic disorders and by allowing them to make informed decisions in planning their family. Prenatal diagnosis can help to identify a genetic disease before birth, giving the family accurate information on which to base decisions concerning that pregnancy.

Presymptomatic therapy refers to the kind of treatment made possible by newborn screening programs, which try to prevent genetic disease processes from progressing postnatally. Prenatal therapy attempts to manage a genetic disease at an exceedingly early stage and can thus be considered a logical extension of presymptomatic treatment.

In treatment of inherited disorders, a specialist often provides medical care for a particular organ system affected by a disease, e.g., pulmonologists and gastroenterologists treat children with cystic fibrosis; hematologists care for patients with sickle cell disease and hemophilia; neurologists care for patients with muscular dystrophies; and nephrologists and transplant surgeons manage patients with cystinosis. Such disorders are representative of multifarious genetic diseases whose management is primarily in the province of subspecialists other than medical geneticists. In contrast, inherited metabolic disorders are generally treated by medical geneticists, and I shall therefore focus on the principles governing the therapy of such illness.

TABLE 1: Frequencies of Some Inborn Errors of Metabolism

Disorder	Incidence
Cystinuria	1:7,000
PKU	1:11,500 (Worldwide)
Histidinemia	1:24,000
Hartnup's disease	1:26,000
Galactosemia	1:75,000
Maple syrup urine disease	1:200,000
Homocystinuria	1:220,000

SOURCE: Adapted from H. Levy, "Genetic Screening for Inborn Errors of Metabolism," *Advances in Human Genetics* (1973):1. Printed with permission.

Inborn errors of metabolism are rare, single-gene disorders most often inherited as autosomal recessive traits. Representative incidence figures (Table 1) range from 1:7,000 for cystinuria to 1:200,000 for maple syrup urine disease.[1] The names of these diseases often refer to the product that is accumulated in physiologic fluids (e.g., galactosemia, phenylketonuria [PKU]) or, in some instances, are names (e.g., maple syrup urine disease) of a more graphic nature. The diseases are the result of defective processing, disposition, or transport of small molecules, e.g., amino acids, fatty acids, metals, or sugars. Such a process is illustrated in Figure 1.[2]

FIGURE 1: Biochemical Sequence: General

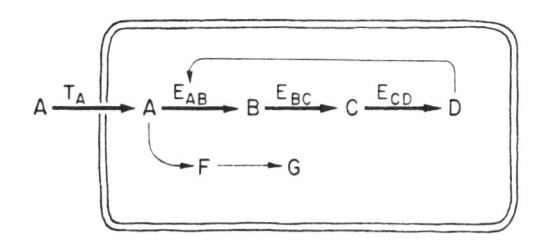

$$A \xrightarrow{T_A} A \xrightarrow{E_{AB}} B \xrightarrow{E_{BC}} C \xrightarrow{E_{CD}} D$$
$$\searrow F \longrightarrow G$$

A, B, C, D — Substrate and Products of Major Pathway
F, G — Products of Minor Pathway
T_A — Transport System for A
E_{AB}, E_{BC}, E_{CD} — Enzymes Catalyzing Conversion of A to B, B to C, and C to D
∥ — Cell Membrane

In the general biochemical sequence of an inborn error of metabolism, substrate A is a small molecule that is transported into a cell by a specific mediated transport system, T_A. In a major processing pathway, substrate A is converted to intermediates B and C and to final product D. These reactions are catalyzed by enzymes E_{AB}, E_{BC}, and E_{CD}. In a minor pathway, A can be converted to F and G. Note that the flux through A→D is far greater than that through A→G. All reaction steps in the sequences, including the transport step, involve proteins and are genetically determined. Inherited disorders of the sequence may exist when there is insufficient quantity or aberrant structure and function of the proteins mediating the various steps.

SOURCE: Adapted from L. E. Rosenberg, "Inborn errors of metabolism," in Diseases of Metabolism, eds. P. Bondy and L. E. Rosenberg (Philadelphia: W. B. Saunders, 1974), pp. 31ff. Printed with permission.

Potential sources of toxicity can be understood from Figure 1. If the conversion of substrate A to product D is inadequate, there will be a dearth of product D, possibly causing disease manifestations. Alternatively, a block in A→B (low activity of the enzyme E_{AB}) will result in an intracellular or extracellular accumulation of A, which may be toxic and cause disease. Further, the flux through A→G may be increased with a block at A→B, and either F or G, normally present at low concentrations, may accumulate to toxic levels. Similar consequences occur if the block is at the transport step.

Such considerations can be used to derive general approaches to therapy, and these are listed below:
- Replacement of product;
- Antagonists;
- Restriction of precursor; and
- Vitamins.

In some instances, therapy consists of replacing a product that is not being synthesized. Examples include administrating clotting factors in patients with hemophilia and thyroid hormone in patients with heritable disorders of thyroid hormonogenesis. In other instances, it may be possible to provide an antagonist to eliminate or otherwise detoxify offending accumulated metabolites. An example is the use of penicillamine as a copper chelator in Wilson's disease. That same drug is also used to increase the solubility of urinary cystine, thereby preventing urolithiasis in patients with cystinuria.

Therapeutic efforts in inborn errors are often directed toward reducing a concentration of a precursor proximal to a block, e.g., substrate A if the activity of enzyme E_{AB} is deficient (See Figure 1). This approach is valid whether toxicity is due to A directly or to derivatives thereof and presently constitutes a major fraction of physicians' therapeutic efforts. Concentrations can be reduced by restricting a specific precursor, e.g., a particular sugar or amino acid, or a class of nutrient, e.g., protein or carbohydrate. Diseases that require nutritional intervention include phenylketonuria, citrullinemia, and maple syrup urine disease: they are discussed below as examples of the medical profession's capabilities and limitations in managing inborn errors of metabolism.

PKU

The clinical course of a phenylketonuric baby recently ascertained in a California newborn screening program is as follows: At age 2 days, the child was discharged from the hospital, and a mandatory screening test for PKU, galactosemia, and hypothyroidism was obtained. The level of 6.1 mg/dl (normal screen <4.3 mg/dl) was reported on day 4, at which time a repeat screening test was obtained. The repeat screening determination (reported on day 6) was 13 mg/dl, suggesting PKU. A confirmatory blood test (direct measurement of serum phenylalanine and tyrosine) was obtained on day 6, and dietary phenylalanine restriction was begun. Treatment was therefore begun during the first week of life.

In the treatment of such children, maintenance of blood phenylalanine in a range between 3 to 10 mg/dl is an acceptable goal, permitting normal growth and development.[3] The blood level has been so maintained in this particular girl through age 1½ years; her physical examinations have been entirely normal; and her development has been excellent in every parameter.

What would have happened if a newborn screening program did not exist and this child had gone untreated? During the first months of life there might have been few symptoms other than a skin rash or dry skin and pigmentation lighter than that of the family. Between 3 and 6 months, development would be noticeably delayed, with neurological abnormalities clearly expressed by 1 year. Microcephaly, electroencephalogram abnormalities, and frank seizures might be observed, and, from retrospective analyses, it is apparent that babies with PKU lose approximately 40 intelligence quotient (IQ) points during the first year of life.[4] This inexorable and tragic decline was prevented in the patient discussed above and in patients similarly treated. However, the paucity of physical findings in PKU and other inborn errors of metabolism seriously compromises physicians' ability to make an early or presymptomatic diagnosis, and it is only through a newborn screening program that physicians are able to ascertain whether a child is affected before the advent of irreversible neurological damage.

FIGURE 2: Phenylalanine Hydroxylase Reaction

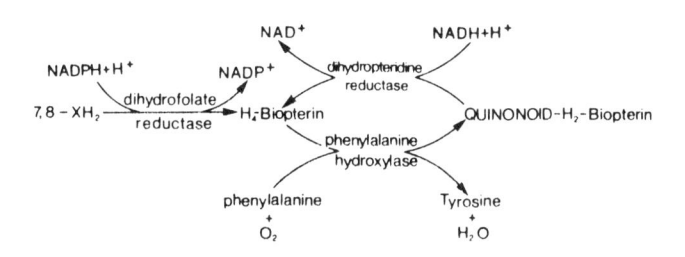

Scheme of phenylalanine hydroxylation and cofactor transformations. The coenzyme ($7,8$-XH_2) as isolated from the liver is reduced by an NADPH-dependent dihydrofolate reductase. The reduced H_4-biopterin participates with L-phenylalanine and O_2 in the phenylalanine hydroxylase reaction. The oxidized cofactor (quinonoid-H_2-biopterin) is then regenerated by an NADH-dependent dihydropteridine reductase, without any further need for the action of dihydrofolate reductase.

SOURCE: Adapted from A. Tourian and J. Sidbury, "Phenylketonuria," in *Metabolic Basis of Inherited Disease,* eds. J. Stanbury, J. Wyngaarden, and D. Fredrickson (New York: McGraw-Hill, 1978). Printed with permission.

The rationale for the treatment of PKU patients can be derived from a consideration of the phenylalanine hydroxylase reaction, depicted schematically in Figure 2.[5] Given the ineffective conversion of phenylalanine to tyrosine, phenylalanine accumulates behind this blocked enzymatic reaction step, and it or derivatives, e.g., phenylketones, are directly or indirectly toxic. The mechanisms of toxicity are unknown, but it has been established that restriction of phenylalanine, i.e., restriction of precursor, permits normal growth and development in classic PKU.

Other potential defects in the reaction step are illustrated in Figure 2 and are related ultimately to a deficiency of the cofactor, tetrahydrobiopterin. Such patients will have hyperphenylalaninemia and PKU, be discovered in screening programs and begun on phenylalanine-restricted diets as if they had classic (phenylalanine hydroxylase deficiency) PKU. Neurological deterioration will supervene despite adequate control of the hyperphenylalaninemia, and other modalities of treatment must be introduced for such patients. This is one of many instances in which a number of different enzyme defects can lead to a similar chemical and similar clinical phenotype. Knowledge of the metabolism of the substrate, then, not only can provide a rationale for treatment protocols, but also can suggest possible reasons for failure of treatment. In disorders of tetrahydrobiopterin generation, the therapy is quite different from that in classical PKU, and involves product replacement rather than restriction of precursor.[6]

The effectiveness of phenylalanine restriction in classic PKU has been demonstrated in numerous studies revealing significant IQ and developmental differences between early- and late-treated patients. A representative example is shown in Table 2.[7] In an extensive project of this nature (the Los Angeles Children's Hospital PKU Collaborative Study), it was observed that the mean IQ of treated children with PKU was slightly but significantly lower than that of their unaffected siblings.[8] Further, the mean IQ in their combined patient sample was lower than the normative mean of 100.[9] Their data suggest the presence of some intellectual impairment, even when treatment was begun early and properly maintained. Explanations for such residual deficits have focused on the need for further refinements in dietary therapy, on the possibility of unknown fundamental conceptual errors in therapy and an incomplete understanding of the pathophysiology of the disease; and on prenatal damage not addressed even with presymptomatic intervention. It is important to realize that even in this best characterized of inborn errors of metabolism,[10] the treatment is not perfect.

In most other inborn errors amenable to nutritional manipulation the empirical data base is not nearly as extensive as in PKU. Nutritional restrictions in such instances may well be deleterious in ways that are not com-

TABLE 2: Comparison of IQ (at age 2 years) In Sibling Pairs Diagnosed at Different Ages But Similarly Adequately Treated For Classic PKU

Sibling Pair	IQ Late-treated Sibling*	Early-treated Sibling†
1	45	93
2	22	113
3	75	92
4	27	100
5	75	100
6	37	70
7	65	88
8	69	86
9	51	94
10	54	96
	Mean IQ 53	93

*Onset of treatment at mean age 11.7 months.
†Onset of treatment at mean age 18 days.

SOURCE: Adapted from C. R. Scriver and L. E. Rosenberg, *Amino Acid Metabolism and Its Disorders* (Philadelphia: W. B. Saunders, 1973).

pletely understood. Therefore, in genetic counseling of parents and discussions of prognosis, optimism may be firmly communicated but must be appropriately tempered with an honest assessment of the uncertainties extant in such therapeutic endeavors. A cautionary note in counseling is especially appropriate when the approach described for PKU is applied to postsymptomatic intervention in other disorders. Two examples follow.

Citrullinemia

The patient arrived at the University of California, San Francisco, during his second week of life after a period of lethargy and hypotonia that had progressed to coma and unresponsiveness to pain. The diagnosis of citrullinemia was made shortly after admission. This disorder is caused by a deficiency of argininosuccinic acid synthetase activity, and such babies cannot adequately detoxify excess nitrogen (ammonia) by conversion to urea. Following immediate procedures to lower the concentration of blood ammonia to normal levels, longer-term therapy has included various modalities (supplemental arginine, protein restriction, supplemental ketoacid analogues of essential amino acids, sodium benzoate) designed to facilitate restriction or elimination of precursor.[11, 12]

In retrospect, it became clear that the initial insult produced severe irreversible brain damage. The infant's blood ammonia elevations were controlled, but retardation was not prevented. In general, postsymptomatic treatment of inborn errors of metabolism is not always successful, and the outcome frequently depends on the severity of the initial insult. The possibility that this patient might ultimately exhibit retardation of unknown degree was considered with the parents, and the decision to treat vigorously was made in conjunction with the family. At age 1 to 2 years the child showed growth retardation, a small head circumference, and delayed development.

Maple Syrup Urine Disease

The child presented at age 9 months with a history of failure to thrive, multiple formula changes, irritability, and developmental delay. He was not able to sit alone. He was admitted to hospital because of severe ketoacidosis and lethargy that had progressed to coma. The diagnosis of intermediate-type maple syrup urine disease was made, and appropriate therapy was instituted. The mother understood that the dietary therapy (restrictions of branched-chain amino acids) initiated at this late age would eliminate the failure to thrive, irritability, formula intolerances, and the episodic neurological dysfunction; less clear was the ultimate extent of the developmental delay. Although the child is showing remarkable progress in growth and development, mild delays in some developmental parameters exist. This case again illustrates that physicians can manage to control chemical derangements reasonably well but cannot reverse brain damage that has already occurred.*

Vitamin-responsive Disease

Vitamin-responsive, or vitamin-dependent, disease, in contrast to the disorders discussed above, constitutes a category of inborn errors in which postsymptomatic treatment has been quite successful.[13] The basis for such disorders is illustrated in Figure 3 and discussed below.

Normal catalysis by enzymes (see Figure 1) may require binding of cofactors to the enzymes. Examples of cofactors include such vitamins as cobalamin, pyridoxine, thiamine, and biotin. Vitamins are themselves small molecules and must be transported to and into cells, reach the appropriate cell compartment, and perhaps be converted to other chemical forms before attachment to the enzyme protein. The processing of vitamins as small mol-

---

*We have recently documented normal growth and developmental parameters in this child at age 42 months.

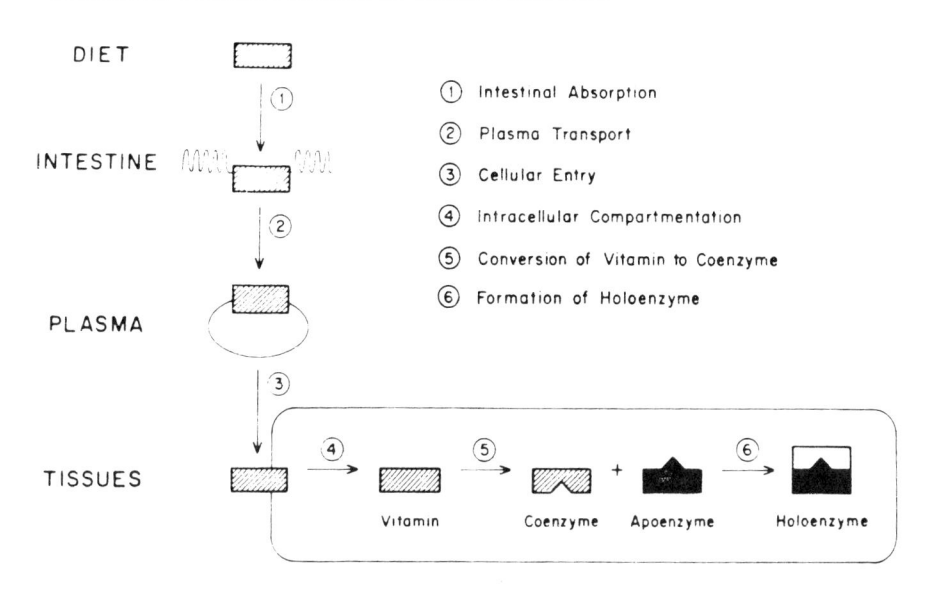

FIGURE 3: Potential Mutations in Vitamin Metabolism

① Intestinal Absorption
② Plasma Transport
③ Cellular Entry
④ Intracellular Compartmentation
⑤ Conversion of Vitamin to Coenzyme
⑥ Formation of Holoenzyme

SOURCE: Adapted from L. E. Rosenberg, "Vitamin-responsive Inherited Metabolic Disorders," *Advances in Human Genetics*, 6(1976): 1-74. Printed with permission.

ecules is not unlike the sequence illustrated in Figure 1 and is elaborated further in Figure 3, which illustrates each of the above steps. One can anticipate that an inborn error of metabolism may affect one of these steps for a given vitamin or may affect an enzyme protein so that it binds its cofactor poorly. In either instance, the result would be defective catalysis by the vitamin-dependent enzyme.

In many instances such blocks in vitamin processing or in coenzyme-enzyme interaction can be significantly overcome if very high concentrations of the vitamin are presented to the reaction system in question. In clinical practice, this is done by administering pharmacological doses of the specific vitamin.

Table 3 lists the disparate clinical findings in two patients with biotin-responsive multiple carboxylase deficiency.[14] The blood and urine chemistries in each instance led physicians to conclude that there was deficient activity *in vivo* of three different carboxylase enzymes. Carboxylases use biotin as a cofactor, and, based on this point and numerous clinical precedents,[15, 16, 17, 18, 19] the physicians administered approximately 250 to 1000 times the usual daily intake of this vitamin to each child. The chemical

TABLE 3: Multiple Carboxylase Deficiency

Neonatal Form
    Tachypnea
    Hypertonia
    Lethargy
    Lactic acidosis
    Ketonuria
    Mild hyperammonemia
    Hypoglycemia

Infantile Form
    Erythematous rash
    Delayed developmental milestones
    Nystagmus
    Candids dermatitis
    Alopecia
    Truncal ataxia
    Keratoconjunctivitis
    Lactic acidosis
    Intermittent ketonuria
    Mild hyperammonemia

SOURCE: Adapted from S. Packman, N. Caswell, and H. Baker, "Biochemical Evidence for Diverse Etiologies in Biotin Responsive Multiple Carboxylase Deficiency," *Biochemical Genetics,* 20, (1982): 19-28.

derangements rapidly normalized, accompanied by dramatic clinical improvement. Both babies are growing and developing normally at ages 3½ and 5 years, respectively. Evidence supports the contention that the patient with the infantile form suffers from a defect in biotin absorption while the baby with the neonatal form suffers from an abnormality in intracellular processing of biotin.[20, 21] In other situations in which chemical reaction steps depend on vitamins, a similar clinical approach can be, and has been, used and has often been successful.[22] Administration of high doses of necessary vitamins restores a significant fraction of biochemical activity *in vivo,* and the patients generally do well, often on unrestricted diets.

In the case of the child with the neonatal onset form, the diagnosis was made by enzymatic studies in cultured skin fibroblasts.[23] When the child's mother became pregnant with a second child, physicians took advantage of the clinical biotin responsiveness and of the biochemical expression of the disease in fibroblasts to formulate a strategy of prenatal diagnosis and prenatal treatment of the fetus. It was their intention, if the fetus were affected, to give biotin in high doses to the mother, knowing that it would be transmitted across the placenta to the fetus. The fetus would then have high serum and tissue biotin levels, which would prevent possible prenatal damage and severe perinatal lactic acidosis (see Table 3).

Based on studies of amniotic fluid cells, the physicians involved learned that the fetus was affected. The mother was given biotin at 23⅓ weeks of gestation, a point in pregnancy well past the major period of organogenesis and teratogenicity risk. The child had no clinical or chemical abnormalities in the neonatal period, and the baby (a girl) has grown and developed normally while receiving continuous postnatal biotin supplements.[24]

In summary, I have reviewed certain capabilities and limitations of treatment approaches in inborn errors of metabolism, and have commented on the timing of the application. Physicians' goals in treating inborn errors of metabolism are to improve and expand the therapeutic repertory and to apply these methods as early as possible during growth and development of the patient. If we consider that presymptomatic therapeutic intervention is an accepted and integral component of the management of heritable metabolic disease, then one can view prenatal therapy as a logical and reasonable extension of such an approach to patient care. I expect that there will be an increasing number of reports of prenatal therapy of genetic disease, as the precision of prenatal diagnosis continues to improve, and as more attention is given to this unique kind of opportunity to prevent disease manifestations in children.

Footnotes

1. H.L. Levy, "Genetic Screening for Inborn Errors of Metabolism," *Advances in Human Genetics* 4 (1973): 1.
2. L.E. Rosenberg, "Inborn errors of metabolism," in *Diseases of Metabolism*, P. Bondy and L.E. Rosenberg, eds. (Philadelphia: W.B. Saunders, 1974), pp 31ff.
3. J.C. Dobson, M.L. Williamson, C. Azen, and R. Koch, "Intellectual Assessment of 111 Four-Year-Old Children with Phenylketonuria," *Pediatrics* 60 (1977): 822-827.
4. C.R. Scriver and L.E. Rosenberg, *Amino Acid Metabolism and Its Disorders*, (Philadelphia: W.B. Saunders, 1973).
5. A. Tourian and J. Sidbury, "Phenylketonuria," in *Metabolic Basis of Inherited Disease*, J. Stanbury, J. Wyngaarden, and D. Fredrickson, eds. (New York: McGraw-Hill Book Co., 1978).
6. C.R. Scriver and C.L. Clow, "Phenylketonuria: Epitome of Human Biochemicals Genetics," *New England Journal of Medicine* 303 (1980): 1336-1342, 1394-1400.
7. C.R. Scriver and L.E. Rosenberg, *Amino Acid Metabolism*.
8. J.C. Dobson, E. Kushida, M. Williamson, E. Friedman, "Intellectual Performance of 36 Phenylketonuria Patients and Their Nonaffected Siblings," *Pediatrics* 58 (1976): 53-58.
9. Dobson, Williamson, Azen, and Koch, "Intellectual Assessment."

10. Scriver and Clow, "Phenylketonuria."
11. M.L. Batshaw and S.W. Brusilow, "Treatment of Hyperammonemia Coma Caused by Inborn Errors of Urea Synthesis," *Journal of Pediatrics* 97 (1980): 893-900.
12. J. Thoene, M. Batshaw, E. Spector, et al., "Neonatal Citrullinemia: Treatment with Keto-Analogues of Essential Amino Acids," *Journal of Pediatrics* 90 (1977): 218-224.
13. L.E. Rosenberg, "Vitamin-responsive Inherited Metabolic Disorders," *Advances in Human Genetics*, 6 (1976): 1-74.
14. S. Packman, N. Caswell, and H. Baker, "Biochemical Evidence for Diverse Etiologies in Biotin Responsive Multiple Carboxylase Deficiency," *Biochemical Genetics* 20 (1982): 17-28.
15. D. Gompertz, G.H. Draffan, J.L. Watts, et al., "Biotin-responsive 1/b Methylcrotonylglycinuria," *Lancet* (1971): 22.
16. L. Sweetman, S. Bates, D. Hull, et al., "Propionyl-CoA Carboxylase Deficiency in a Patient with Biotin-responsive 3-Methylcrotonylglycinuria," *Pediatric Research* 11 (1977): 144.
17. M. Saunders, L. Sweetman, B. Robinson, et al., "Biotin-responsive Organic Aciduria: Multiple Carboxylase Defects and Complementation Studies with Propionic Aciduria in Cultured Fibroblasts," *Journal of Clinical Investigation* 64 (1979): 1695.
18. B. Charles, G. Hosking, A. Green, et al., "Biotin-responsive Alopecia and Developmental Regression," *Lancet* 2 (1979): 118.
19. K. Roth, R. Cohn, J. Yandrasitz, et al., "Beta-Methylcrotonic Aciduria Associated with Lactic Acidosis," *Journal of Pediatrics* 88 (1976): 229.
20. S. Packman, L. Sweetman, M. Yoshino, et al., "Biotin-responsive Multiple Carboxylase Deficiency of Infantile Onset," *Journal of Pediatrics* 99 (1981): 421-423.
21. Packman, Caswell, and Baker, "Biochemical Evidence for Diverse Etiologies."
22. Rosenberg, "Vitamin-responsive Inherited Metabolic Disorders."
23. S. Packman, L. Sweetman, H. Baker, et al., "The Neonatal Form of Biotin-responsive Multiple Carboxylase Deficiency," *Journal of Pediatrics* 99 (1981): 418-420.
24. S. Packman, M. Cowan, M. Golbus, et al., "Prenatal Treatment of Biotin-responsive Multiple Carboxylase Deficiency," *Lancet* 1 (1982): 1435-1439.

chapter 6

An Introduction To Genetic Engineering

H. Michael Shepard, PhD

Genetic engineering has as its goal the construction of organisms with desired traits. Examples of one kind of genetic engineering are the breeding of cattle to yield more meat or milk and the selective cross-hybridization of different varieties of corn to give new hybrids that are more resistant to disease or that have a higher content of protein than either of the parents. Genetic engineering in this sense is therefore not a new branch of biology; rather, it is one of the oldest.

As it has been recently popularized by the media, however, genetic engineering is the branch of molecular biology that deals with the isolation and manipulation of particular genes in the laboratory. Genes that have been isolated and chemically characterized include those which direct the synthesis of hemoglobin, insulin, growth hormone, and interferon. Among the most amazing developments in the field of genetic engineering has been the technology required to clone, i.e., to make multiple identical copies of particular human genes and to then use the cloned genes to direct the synthesis of their proteins in bacteria. This development has subsequently allowed the large-scale production of medically useful proteins that, a few years ago, were virtually unobtainable. The idea of inserting a human gene into a bacterium and then getting the gene to express (produce) its protein in such a foreign environment is astounding. This was a great conceptual accomplishment.

Dr. Shepard is in the Department of Molecular Biology, Genentech, Inc., South San Francisco, CA.

In light of the above, genetic engineering can help scientists to do the following:
- Understand how genes work in their natural environment;
- Produce useful quantities of naturally scarce proteins with applications to medicine or agriculture;
- Diagnose genetic disease; and
- Eventually treat genetic disorders at the level of the gene.

Each of these goals has become approachable only in the last few years, because every animal cell has tens of thousands of genes. Each gene, chemically a segment of the deoxyribonucleic acid (DNA) molecule, therefore makes up only an infinitesimal proportion of the total genetic information contained in each cell. To approach this problem, the techniques of gene cloning were developed. Essentially, cloning is a complex series of experiments that result in the placement of a single eukaryotic gene into a microorganism, usually the bacterium *Escherichia coli*. The result of this procedure is that the gene, which comprised less than a ten-thousandth of the genetic information in the natural host cell, will now comprise as much as one-tenth of the DNA in *E. coli*. This amplification is possible for two reasons. First, the number of genes in *E. coli* is a thousandfold less than in the eukaryotic cell from which the gene was taken. Second, as shown in Figure 1, *E. coli* contains circular DNA structures called *plasmids* that exist in

---

Figure 1. Conversion of genetic information into protein. This figure demonstrates three of the major processes of any living thing. These include the *replication* of genetic information contained in the circular plasmids (top of the figure), which ensures that the living entity, as it has evolved to the present time, will not change drastically between generations. Immediately after a plasmid is introduced into the bacterium, *E. coli*, it replicates many times. The chromosome of *E. coli* replicates only once in each generation. Also shown is *transcription*, which is the process by which information in storage form (DNA) is changed into messenger ribonucleic acid (mRNA). mRNA is used to program the next step, *translation*. Translation, like replication and transcription, is a transformation of information. In translation, cellular components known as *ribosomes* play a major role. Essentially, their role is to bring the building blocks of proteins (amino acids) together so that they can be polymerized into protein. The amino acids are brought to the ribosome attached to structures called aminoacyl-transfer RNAs (the cloverleaf structures shown above the boldface arrow). The three amino acids shown are formylated methionine (f-Met), glutamine (Glu), and tryptophan (Trp). The *AUG* sequence of ribonucleotides at the left end of the mRNA in the figure comprises the start signal. The *UAA* sequence at the other end is the stop signal. Ribosomes attach to the mRNA at the *AUG* sequence, begin synthesizing a specific protein as the mRNA dictates by the sequence of the three nucleotides, and then falls off when it "reads" the *UAA* sequence. More details of the process are shown in Figure 2. Both figures adapted from: R. Wetzel, "Applications of Recombinant DNA Technology." *American Scientist* 68 (1980): 664-665. Reprinted with permission.

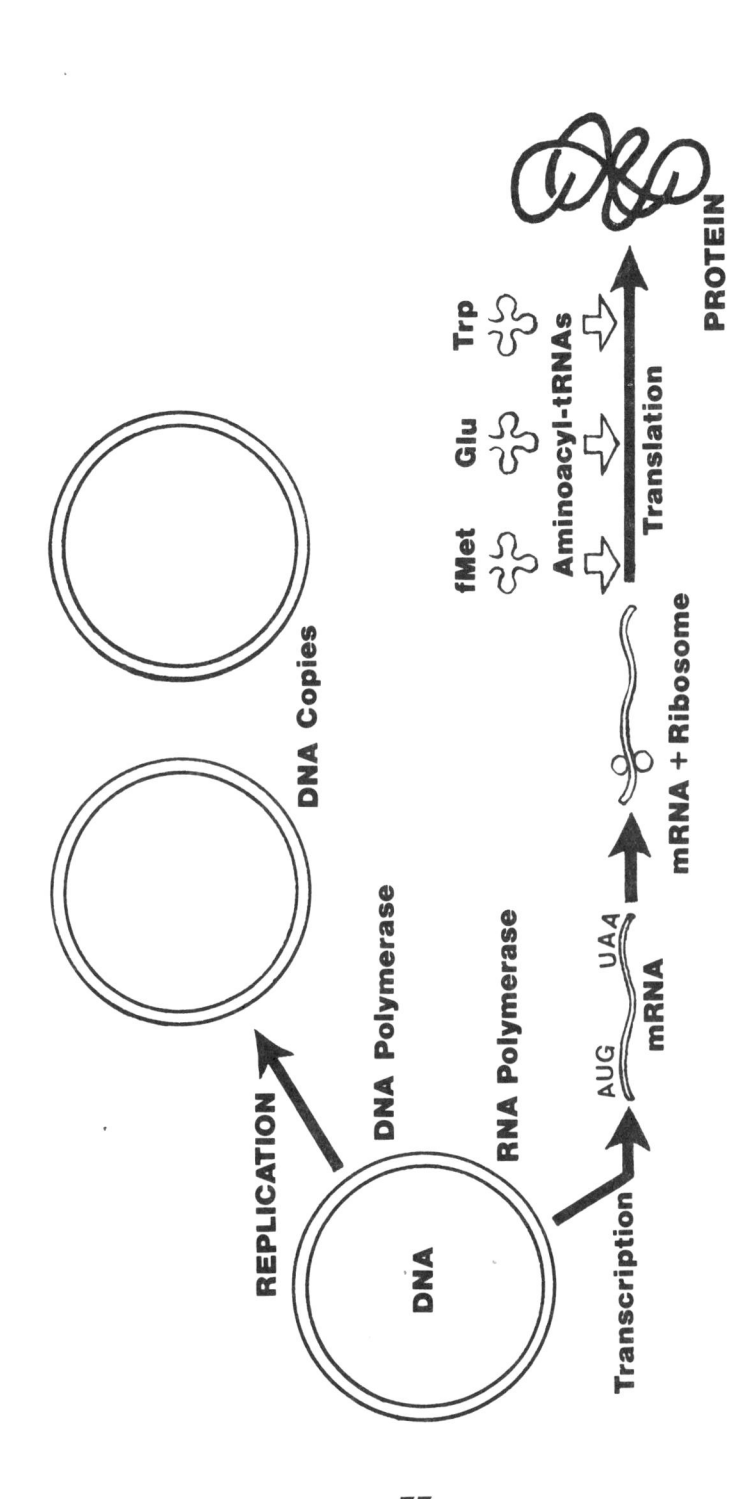

FIGURE 1. Conversion of Genetic Information into Protein

many copies in each cell. The cloned gene is placed into these plasmid carriers before it is put into *E. coli.* As a result, when the plasmid replicates in *E. coli,* so does the cloned gene. This amplification then allows scientists to isolate large quantities of the gene and consequently to examine its structure and begin studies with the aim of understanding how the expression of the gene is controlled in its natural environment.

Cloning: Procedure and Implications

As previously mentioned, all genes are made from DNA. DNA is a linear arrangement of four deoxyribonucleotides: adenine (A), cytosine (C), guanine (G), and thymine (T). The DNA is a very long, threadlike molecule in the shape of a double helix. (A single helix is like a spiral; a double helix is two intertwined spirals, somewhat akin to a ladder twisted into a spiral shape.) The molecule is actually two different spiral strands of DNA bound together by bridges, because these bases, like C and G, stick together; they are "complementary." The same holds true for the other bases, A and T. Consequently, the two intertwined spiral strands are linked together along the whole length of the molecule (Figure 2). What actually makes a gene is the particular order in which these bases are arranged in DNA. This arrangement actually dictates the kind of proteins that are made; Figure 1 indicates how this process is carried out. In any biological system, DNA is replicated by the cell to pass information to the next generation, where it directs the synthesis of new protein by two sorts of operations (Figure 1). The first step is the production of messenger ribonucleic acid (mRNA) from DNA by the process of transcription. mRNA is so called because it brings the genetic message from the gene to the protein synthetic machinery of the cell (the ribosomes), where the information is translated into protein. During the transcription of DNA into mRNA, the thymine component of DNA is replaced by the base uracil (U) of RNA.

In cells, usually only DNA is replicated and passed from one generation to the next. To clone a particular gene scientists usually start from mRNA, because genes in the DNA molecule are very complicated. They contain a number of signals called "introns" which interrupt those portions of the gene which actually direct the synthesis of protein. The mRNA molecule is simpler because it has been processed to contain the information that makes the protein without any interrupting signals. Hence, scientists first make the mRNA, which requires isolating it from the cell. But to propagate it in *E. coli* the information must first be converted back into a DNA molecule. This is done with the enzyme "reverse transcriptase," which transfers the information on mRNA back into a DNA molecule. Once this is accomplished the

FIGURE 2. The Genetic Code

Complementary 5'─∥─ ATGGAAATTATACGTATGCCTGAAGAGTAA ─∥─ 3'
Strands   3'─∥─ TACCTTTAATATGCATACGGACTTCTCATT ─∥─ 5' DNA

↓ (Transcription)

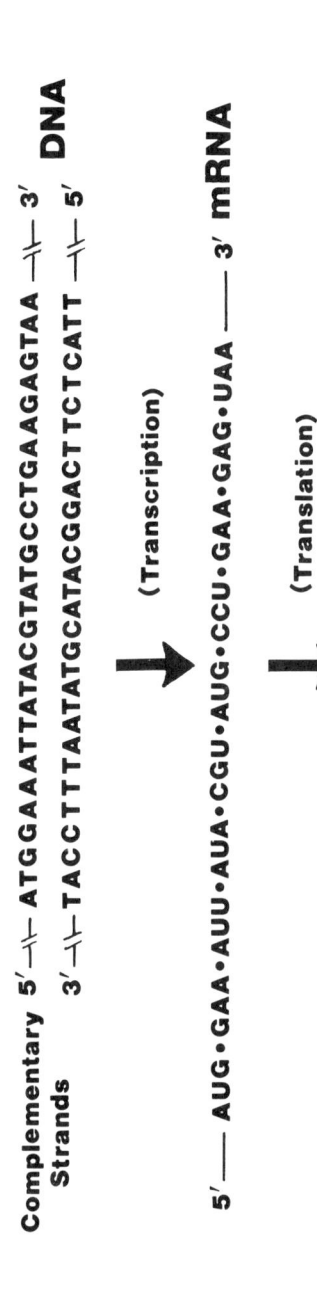

5'── AUG·GAA·AUU·AUA·CGU·AUG·CCU·GAA·GAG·UAA ── 3' mRNA

↓ (Translation)

fMet–Glu–Ile–Ile–Arg–Met–Pro–Glu–Glu ^stop PROTEIN

Figure 2. The genetic code. Each "molecule" of DNA in a cell is actually comprised of two complementary strands of DNA. In DNA, the bases A and G are complementary to the bases T and C, respectively; mRNA comes from only one strand (the one that is oriented, by convention, 3' to 5'). Note that in DNA and RNA the nucleotide usage is different. That is, the T of DNA becomes the U of RNA. The mRNA sequence is complementary to the DNA strand from which it is synthesized. In other words, it is the same information conveyed in a slightly different language. Thus, it has the same nucleotide sequence as the nonactive DNA strand (except for the U and T difference). Each group of three nucleotides is called a codon and encodes a single amino acid (e.g., *AUG*-methionine), except for the *UAA* sequence, which encodes no amino acid and indicates termination of synthesis.

new DNA molecule, which does not contain the introns of the gene in the original host cell, can be spliced into a plasmid and introduced into *E. coli*. The plasmid can then replicate to a high number of copies inside the bacterium and produce large quantities of its protein. This process as it was utilized in the cloning of human immune interferon is shown in Figure 3.

What is then done with the selected DNA that is present in very large quantities? In the beginning, the first goal of genetic engineering was to clone a gene and then study its structure, because methods now exist for examining the whole gene, i.e., the DNA sequence that makes up the gene. If one looks at a number of these genes and their structures, one can rather quickly find and know those parts that the genes have in common. This suggests that those particular nucleotide sequences may be important for regulation.

DNA sequencing (i.e., determining the sequence of nucleotides in a DNA segment) is a very fast procedure compared to obtaining amino acid sequence data. Hence, if one wants to learn the amino acid makeup of a protein, it is much easier to look at the gene that encodes the sequence of amino acids of that protein than it is to try to determine directly the sequence of amino acids in the protein, which is difficult to purify chemically and to sequence. Organic synthesis techniques have been recently developed which allow us to easily change these coding sequences and then make proteins with altered properties.

Prenatal Diagnosis of Genetic Disease

Cloned DNA segments can be used to diagnose genetic disease prenatally. Genetic diseases result from a spontaneous and deleterious change in the structure of a gene whose protein product is required for the growth or maintenance of the organism. It is important to note that prenatal diagnosis by molecular biological techniques requires some risk (although small) to

---

Figure 3. The construction of a gene library. The steps involved in the isolation of human immune interferon messenger RNA, its conversion back into DNA and finally its insertion into a plasmid carrier are shown. The DNA copy of messenger RNA is hybridized to the plasmid vector after a string of "C" residues is added to the DNA copy and "G" residues are added to the plasmid. As described in the text, "C" residues hybridize to "G" residues thereby allowing the plasmid to recircularize. The plasmid circles can enter the *E. coli* host, where each replicates several times giving a final copy number of about 30 plasmids per cell. The transformed cells are spread onto agar dishes containing tetracycline to select for those cells containing plasmids. Adapted from: H.A. de Boer and H.M. Shepard. Strategies for optimizing foreign gene expression in *E. coli*. *Horizons in Biochemistry and Biophysics*, in press. Reprinted with permission.

FIGURE 3. Construction of a Gene Library

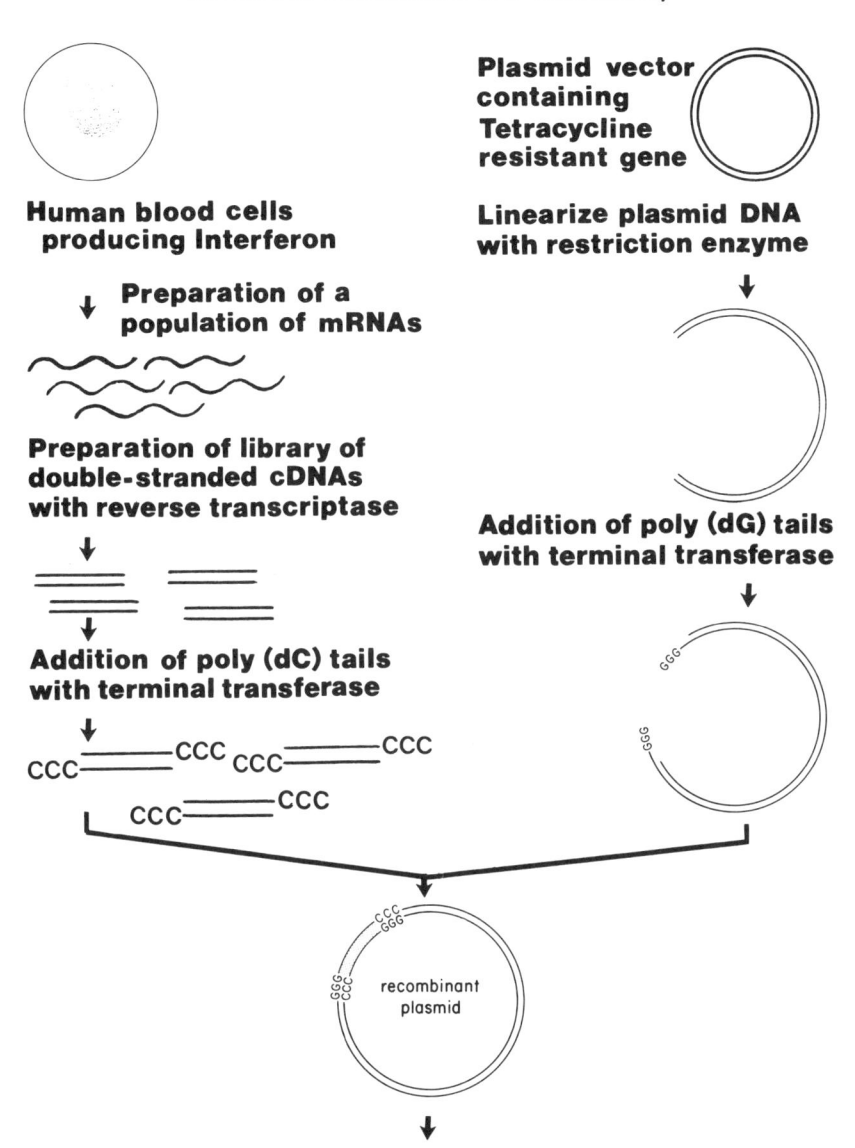

Human blood cells producing Interferon

↓ **Preparation of a population of mRNAs**

Preparation of library of double-stranded cDNAs with reverse transcriptase

↓

Addition of poly (dC) tails with terminal transferase

Plasmid vector containing Tetracycline resistant gene

Linearize plasmid DNA with restriction enzyme

↓

Addition of poly (dG) tails with terminal transferase

↓

recombinant plasmid

↓

Transform E. Coli sensitive to tetracycline, select tetracycline resistant clones and screen bacteria containing recombinant plasmids for interferon gene.

the fetus. For this reason, such techniques (e.g., amniocentesis) should be applied only in families in which genetic counseling has shown predisposition toward the disease.

One very interesting application of genetic engineering to the field of prenatal diagnosis has been detection of defects in the growth hormone gene of a fetus. Using some of the techniques of molecular biology, it is possible to display all the genes of a given person according to size in a gel matrix. In this fashion, the longer genes are trapped at the top of the gel matrix while the smaller ones tend to be toward the bottom. There are so many genes, however, that no particular ones can be distinguished visually by this technique. Normal growth hormone genes migrate to a particular place in such gels. If a deletion occurred in the gene (absence of a part of the DNA segment), then it would migrate to a place characteristic of a smaller piece of DNA. With a technique called hybridization, it is possible to detect a particular gene among the huge number of genes found in a human cell.

In hybridization, a cloned human growth hormone gene is made into a radioactive probe. It is then tested with the DNA from the gel matrix described above. The probe will only stick to other growth hormone DNA in the gel. After the gel is hybridized and washed, it is exposed to x-ray film. Wherever the probe was hybridized with DNA from the gel matrix, a radioactive spot is seen on the x-ray film. The bands obtained from a healthy person differ from those of a person with, for instance, genetic dwarfism (Figure 4). This experiment can also be done with cells in the amniotic fluid of a fetus. An example of this type of dwarfism is found in Switzerland. Because of the country's numerous mountains and valleys, people in Switzerland are relatively isolated, one population from another. As a result, if a particular gene mutation occurs in one valley, the gene is propagated in that population because the people tend to marry one another rather than go over the mountains to another valley. Consequently, a particular kind of genetic dwarfism occurs in certain locations in Switzerland and is the result of a gene mutation.

Because all cells of the body, except the reproductive cells, have two copies of a gene, if a mutation occurs in one copy the person usually does not exhibit any symptoms unless the disorder is due to a dominant gene. If the mutation occurs in both copies of the gene, symptoms of the disorder will generally be noted. Young children who have a double mutation of this particular gene do not make any growth hormone at all. Such individuals are genetic dwarfs. Yet because they make no growth hormone at all, they cannot be treated with growth hormone. The reason for this paradox is that to these people growth hormone is a foreign protein and they will mount an immune response to it. They will make antibodies to any human growth

FIGURE 4. IGHD-IA Restriction Analysis

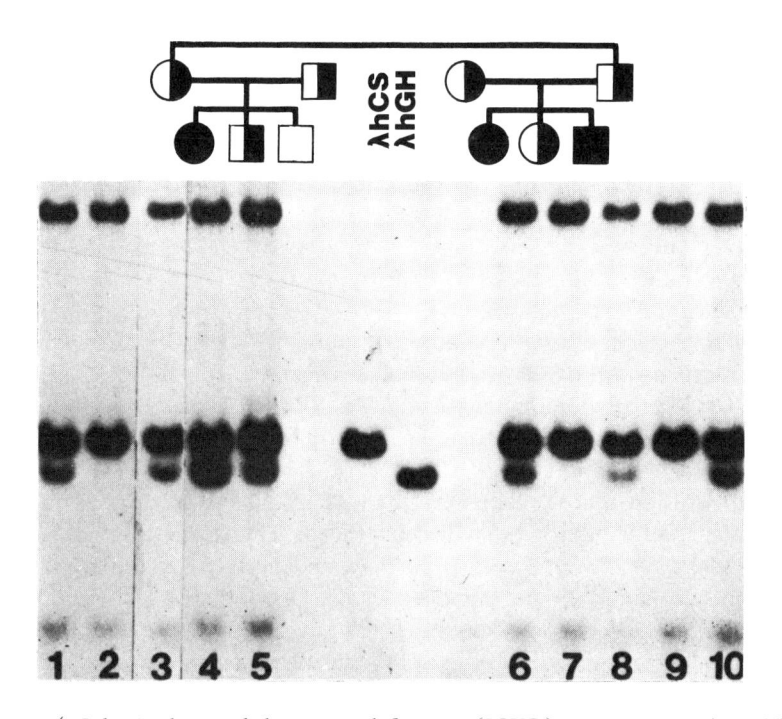

Figure 4. Inherited growth hormone deficiency (IGHD) restriction analysis. This figure portrays one of the ways in which a molecular biologist looks at genetic disease. This is an example of an analysis done with one family (1A). By convention, circles represent females and boxes males. Except for genes carried on the X and Y chromosomes of the male, there are two copies of every gene in each human cell, one from the father and the other from the mother. A filled-in circle-square represents a person who has no normal growth hormone genes (homozygous, mutant), while an empty circle-square represents a person who is totally normal (homozygous, wild type). The half-filled circle-square stands for the person having only one normal growth hormone gene. Although there is a considerable amount of information in this figure, only the spots indicated by the arrows should be considered. Note that people who have normal growth hormone genes on both chromosomes also completely have DNA indicated by the arrows. Those with one normal and one mutant gene (heterozygotes) have a smaller amount of DNA in the location indicated by the arrow. This is difficult to see in the figure, but is most apparent in sample 8. Those with two copies of the mutant gene have no growth hormone DNA in the smaller-sized gene fragment. Source: J.A. Phillips, III, B.L. Hjelle, P.H. Seeburg, *et al.* "Molecular Basis for Familial Isolated Growth Hormone Deficiency." *Proceedings of the National Academy of Sciences* 78: 6372-6375, 1981. Reprinted with permission.

hormone injected into them, and these antibodies will render inactive the growth hormone.

If this disease can be diagnosed prenatally, or very shortly after birth, i.e., before the person's immune system has begun to function, then growth hormone treatment can be effective. Such a person can then grow normally, but treatment must occur before or very soon after birth.

Some forms of sickle cell anemia are also due to deletions in genes that make the oxygen-carrying molecule hemoglobin. Techniques similar to those described above have been used for prenatal diagnosis of this disease. Finally, scientists at the University of California have received much publicity in their use of human subjects in trying to correct the serious disease beta-thalassemia through gene engineering. So far, such efforts have met very limited success in mice but not in humans. There are factors in gene expression that are barely understood at this time. It is difficult to predict when single gene therapy may become possible in humans, but there is no way at present that scientists can change the basic genetic character of any mammalian population.

In conclusion, recombinant DNA technology aims to produce certain proteins that are difficult to obtain by other means. This technology has resulted in unique ways of diagnosing prenatally, and perhaps eventually treating, persons with incurable genetic diseases. The work that has been done thus far in the field of genetic engineering has as its major contribution to medicine the large scale production of biologically important proteins and has also added to our understanding of eukaryotic gene structure and function. Future goals in this field include the production of plants that require less fertilizer and can produce adequate amounts of food under adverse conditions. Medically important proteins and inexpensive vaccines for diseases of humans and agriculturally important animals are additional objectives. Much of these various advances are of special importance to Third World countries because of the frequent prevalence of malnutrition and disease in these areas. Thus, genetic engineering offers a further means by which seemingly esoteric technological discoveries can be used to improve the health of people in everyday situations.

PART III
ETHICAL AND SOCIAL ASPECTS

chapter 7

Ethical Principles and Genetic Medicine

Rev. Donald G. McCarthy, PhD

In this chapter I wish to present a framework and an overview of the ethical aspects of genetic disorders and diagnosis. First, I will reflect briefly on the difference between law and ethics. Second, I will discuss ethical systems in general and ethical principles that are relevant to genetic medicine. Third, I will attempt to draw some applications in a rather schematic way.

Law Versus Ethics

What is the difference between law and ethics? In general, law attempts to order the human community in pursuit of liberty and justice. Thus, Germain Grisez and Joseph Boyle titled their book *Life and Death with Liberty and Justice* when they examined public and legal aspects of the euthanasia debate.[1] The two principles of liberty and justice in some way constitute cornerstones for good law. Society needs equal protection of the law for all its members. This, of course, raises questions about those members who have not yet seen the light of day, i.e., fetuses, but there is also a broader challenge to structure the laws of society to ensure life, liberty, and the pursuit of happiness for all citizens. This is the language of the Declaration of Independence. The preamble to the United States Constitution speaks of pursuing the general welfare. In political theory, that is often described as ensuring conditions for the achievement of the common good. Law can thus be described in terms of the definite role it plays in the individual fulfillment that is made possible to each citizen.

Fr. McCarthy is at St. Cecilia's Parish, Cincinnati, OH, and director of education, Pope John XXIII Medical-Moral Research and Education Center, St. Louis, MO

On the other hand, laws do not deal exhaustively with all kinds of human actions. Laws do not forbid all evil actions or insist on all good actions. In short, there must be limitations to what can be legislated. If a particular action does not undermine equal protection and equal rights of all people and does not harm the general welfare, it need not be prohibited by law. It is not particularly appropriate to pass laws against swearing or blasphemy, or even against lying, although the United States does prohibit lying under oath in court and has laws against perjury. The simple fact is, however, that there is no attempt to make everything immoral or unethical to be illegal as well.

Some laws that are enacted, however, are not in accord with good morality or good ethical judgment. Many people still remember the way the law was put to use by the Nazi regime in Germany. Certainly the laws leading to the extermination of the Jews and to the Holocaust were immoral and unjust laws. In this country, laws supporting the practice of slavery were also unjust and immoral. Legality is not a synonym for morality. Many states had laws against interracial marriage, yet in society's present understanding of marriage and human equality, such laws would be considered unjust and immoral.

Those who see a tremendous abdication of the protection of human life in the Supreme Court abortion decisions of 1973 regard those decisions as having undermined what they regarded as just laws protecting the unborn. Those decisions qualify as unjust court decisions in the context of the respect-life ethic in which I am writing.

Ethical Systems and Principles

Ethical Consensus

Ethics can be described in many different ways. A simple descriptive statement would define it as a systematic study of human behavior in order to distinguish good acts from evil ones. By the very fact of being a systematic study, ethics can be studied from a variety of approaches and with a variety of frameworks. Despite the different ethical systems, most people agree on a general ethical consensus, simply because of shared human concerns about what seems morally right or morally wrong.

Among the ancient documents that should get credit for expressing such basic ethical convictions, I would include the Code of Hammurabi and the Ten Commandments of the Judeo-Christian tradition, not only in their negative prohibitions but in the positive precepts they inculcate. Since most of the Commandments are written negatively, one must realize that a positive moral precept is implied in each negative prescription. Thus, "Thou shalt not kill" also commands care for human life. Despite different ethical systems, there is nevertheless a general consensus on some very basic principles of what constitutes a good and fulfilling human existence. The 1973 *Roe v. Wade* decision created a great deal of controversy, but the abortion decisions did not in any direct way destroy what is still a very solid consensus in law: to take the life of innocent persons outside the womb remains a capital crime. That consensus is extremely important in a society that professes to be built on the goals of liberty and justice.

Working Principles for Moral Judgment

Various systems of thought try to organize working principles for moral judgment coherently.[3] One is called *emotivism*. It was popularized during the philosophical movement of linguistic analysis in Great Britain in the 1940s and 1950s, especially by Alfred Jules Ayer.[3] The system really has no cognitive structure of ethical language because ethical statements are ultimately feeling statements. Ayer, following the tradition of British empiricism, rejected the possibility of truth statements that involve ethical values. Wherever a statement was made that included an ethical value, Ayer judged that the statement was not a truth-telling statement, but a feeling-communicating statement. Thus, if someone said, "Abortion is evil," Ayer would translate that as, "Abortion is repulsive to me; abortion makes me feel displeased and unhappy."

Fortunately, I think, not many people try consistently to promote ethics on the basis of emotivism. Since there really can be no ethical system of a cognitive fashion in emotivism, those who subscribe to it simply turn to nonphilosophical sources for their ethical convictions. They may turn to religious traditions, to a broad common consensus, or even to the law as their source for ethical principles.

Another system for organizing ethical principles is *intuitionism*. In this system good and bad are recognized by a nonrational experience, something welling up from within.[4] This differs slightly from emotivism because intuition can be a matter of moral consensus and moral judgment; there is something that rings true in intuition. Some basic convictions are prominent in moral thinking and do arise, in my judgment, from nonreflective processes. I think there is room, however, for organizing intuitions and examining their source, i.e., reflecting on them, clarifying them, purifying them, and even sorting out those that are in some way misleading. This rational process goes beyond the simple system of intuitionism.

Consequentialism

The third system is called *consequentialism*. It is pervasive in much ethical writing and has many variations and different facets. I will describe it broadly by saying that, in it, human beings find out what actions are evil by finding out which actions produce a predominance of evil consequences. This means that human actions themselves cannot be described in an irreversibly evil fashion because one must always look at a given action in its setting and circumstances. A term that one contemporary theologian uses in his discussions of consequentialism is *exceptionism,* because once one has espoused this basic approach, there is always room to make exceptions to general moral rules or moral norms.[5] In this system, new compensating values may always tilt the balance in favor of good effects or good consequences of actions, as opposed to evil.

Another term closely related to consequentialism is *utilitarianism,* which can be traced to philosophers such as John Stuart Mill. In this ethical theory one judges which actions are good according to whether they promote the greatest good of the greatest number of people; this is fundamentally a consequentialist approach.[6] It brings in a social dimension, i.e., the greatest good of the greatest number, but it involves the practical difficulty of estimating that greatest good. It could permit the real possibility of sacrificing individuals or minority groups for that greatest good. One might ask, "Can one defend human rights in the context of the United Nations declaration of human rights in the ethical system of pure utilitarianism?" If people are going to evaluate human actions exclusively in accord with the greatest good of the greatest number, society comes to a kind of Caiaphas principle of sacrificing one or a few for the greatest good of the greatest number.

Obviously, I am oversimplifying this question. This debate has gone on for several hundred years. There are those who find ways of building a modification into consequentialism or utilitarianism to adjust for violations of basic justice or basic human rights. But, for example, in the case of the abortion of the defective fetus, one can find many consequences of that act that seem to be favorable, e.g., delivering parents from a lifelong burden and delivering a child from what could be an unhappy existence, at least in the judgment of others. The model of consequential thinking thus influences some ethicists in the area of genetic concerns, particularly in the abortion of defective fetuses.

In aborting defective fetuses the good of life for one being is sacrificed for the apparent good of others. This points to the problem of weighing the consequences, of measuring the incommensurable, of trying to reach a rational comparison between two different human goods. It seems that the judgment of the greater good will depend on who is defining the greater good. We do, in fact, attempt to make various kinds of cost-benefit analysis in human dilemmas when roughly comparing different kinds of goods. But

Ethical Principles and Genetic Medicine

consequentialism allows directly evil actions that destroy human goods, e.g., the good of life, so that other goods may flourish. Can the moral good of individual life, dignity, and justice be weighed against subsequent desirable effects and long-range consequences?

This brings back the question of the individual in society and how far the individual should be subordinated to the social good. How far should the good of the individual be compromised by principles of social worth? After World War II, Maritain, for example, wrote about the person and the common good because totalitarianism had emerged in modern Europe and precipitated World War II. One concern many people feel is to avoid subordinating persons to society in the way in which society would be the whole and each member a faceless part.

The dignity and meaning of human personhood, both in a secular and in a theistic or theological sense, creates a dynamic tension between the person and society. It refuses to subordinate persons to society in the kind of physical dependence that is very attractive to those who regard society as an organic whole, as, to some degree, the Marxist analysis does. The ethical system of consequentialism, at least in its more extreme forms, seems to undermine the inviolability of the person.

Voluntarism

A fourth kind of ethical system was called *voluntarism* in the late medieval period and, in more modern times, especially since Emmanuel Kant, deontologism.[7] This system says that ethical norms have been given to society and must be obeyed because they are established as the principles of human existence. In some sense, this system is the ethics of command—things are good because they have been commanded. The reverse is more appealing to many in the theist tradition, who prefer to think of God as commanding things because they are good or forbidding things because they are evil. In voluntarism, however, the law, whether divine or human, determines what is good. Those who regard the law as the source of morality are then able to say, for example, that because abortion is legal it is moral. Most people, however, are reluctant to accept such legislative voluntarism in Supreme Court decisions and in the decisions of finite and fallible legislators. Even in the case of God, most people feel that in his wisdom, God has reasons for declaring actions evil or good. Pure voluntarism is thus irrational and unappealing to many.

Natural Moral Law

The fifth and final ethical system is the one that, in classic philosophical thinking of the tradition of scholasticism, was called the *natural moral law*

91

system. The term *natural moral law* has fallen somewhat into disfavor, partly because of new and more dynamic concepts of nature itself. With the development of post-World War II existentialism and an emerging emphasis on personalism, ethicists began to look at the natural moral law in a more personalistic and dynamic sense. Frs. Ashley and O'Rourke chose to designate their approach to this moral system as *prudential personalism* in their book, although Fr. Ashley has also favored the term *transculturalism.*[8]

In natural moral law, human actions are seen in the context of the direction they take people, i.e., in the way they advance or fail to advance human good and human fulfillment and in the way they promote or undermine human well being. Contrary to voluntarism, natural moral law or, in contemporary language, prudential personalism, holds that good human actions *do* have good effects. This system, however, does not always recommend discovering the good of human actions by the consequential method of discovery. It recognizes some basically evil actions that can never be justified by good consequences. Those who accept this position recognize human rights on the basis of the very order of human existence, or the order of creation, if the world is God's world. This system has an a priori conviction that evil actions have long-range evil consequences which cannot be neutralized by a calculus of impressive good consequences.

This system should not be uniquely tied to the scholastic philosophy of the Middle Ages or to Catholic thinking. It is true that it has been very prominent in the Catholic past, at least in pre-Vatican II Catholic theology and philosophy. But it seems clear that the Nuremberg war trials were based on natural moral law principles, which indicates they are not exclusively Catholic or scholastic.

Unfortunately, some people have misunderstood how to apply natural moral law. These principles are not indelibly printed in the minds of all human beings, ripe for applying. There are different levels of certainty and clarity among natural moral law principles. In the language of the Middle Ages, there are primary and secondary precepts of the natural moral law. Furthermore, there is room for a good deal of controversy about exceptions to these laws, which the Catholic Church is currently experiencing with regard to applying natural moral law principles to the use of contraception and sterilization.

The theology of Catholicism recognizes that human beings have a darkened intellect and a weakened will, and therefore, in human, feeble efforts to reach a clear and definitive consensus on natural moral law principles, there will be some difficulties. There will be progress, as in the development of a natural moral law prohibition of slavery, but there will be confusion, as when science offers new technology in practices such as abortion and contraception. I suggest that Jesus gave the spirit of truth to the Church to present the natural moral law in the light of his teaching. Thus, the Church claims this as part of its mission.

Relevant Principles

Certain principles are applicable to genetics. I will discuss them here in the context of the practical life of the Catholic-Christian community and the broader context of the fifth ethical system, that of natural moral law or *prudential personalism.* Within Catholic moral theology since Vatican Council II, continuing reexamination of the natural law principles has been underway. A strong challenge has been mounted against admitting any moral principles as designating specific kinds of actions irreversibly evil or any moral principles as exceptionless. Three of these principles are examined below.

Respect life. The first principle is the moral obligation to respect life and, specifically, the lives of members of the human family. This means it is evil to kill any innocent members of the human family. Those who recognize the unborn to be members of the family would also prohibit their killing. There has been debate on whether fetuses are full members of the human family. Is there some developing stage of human dignity during the 9 months of gestation? One might be inclined to defend that position on intuitional grounds if those in the Catholic community used the ethical system of intuitionism.

But the very strong scientific awareness of the continuity of the growth and development of the fetus leads to a contrary rational conclusion, namely, that the fetus or embryo is a full member of the human family despite its early stage of development. Surely the mere fact of delivery does not in itself contribute to human dignity. A child who has full human dignity on the morning of birth had it the night before in its mother's uterus. Working backward, there is no significant point during the nine months of gestation where human dignity might begin, save at the beginning.

Some Catholic moral theologians are reluctant to speak in the language of the traditional moral prohibition of directly killing the innocent. They prefer proportional moral analysis or proportionalism. This approach simply says that one must not destroy human life without proportionate reason. This could be formulated cautiously by saying that one might only take human life to save human life, and then only as a last resort, provided that such an action did not in the long range undermine the value of human life.

This method of ethical analysis does not focus primarily on the act of taking human life but on the fact that there is no proportionate value for that act. Thus, some would prefer to say that one may not directly take innocent life except to save human life or the moral equivalent of life. In that last phrase, the *moral equivalent,* one opens up possibilities of taking human life to avoid misery or serious burden or to prevent a pregnant woman's loss of sanity, for instance. In this way, proportional analysis can be used to approve both abortion and mercy killing. Authentic Catholic teaching has not used proportionalism, however. The *Declaration on Abortion* in 1974 offered no hint of that estimation of proportionate values, nor did the 1980 *Declaration of Euthanasia* by the Sacred Congregation for the Doctrine of the Faith.

In fact, the latter *Declaration* very explicitly states that the act of mercy killing remains a grave moral evil despite the individual judgment excusing it that might diminish moral culpability:

> It may happen that, by reason of prolonged and barely tolerable pain, for deeply personal or other reasons, people may be led to believe that they can legitimately ask for death or obtain it for others. Although in these cases the guilt of the individual may be reduced or completely absent, nevertheless, the error of judgment into which the conscience falls, perhaps in good faith, does not change the nature of this act of killing, which will always be in itself something to be rejected.[9]

Promote Health. The second principle teaches the moral obligation to promote health and bodily integrity. This principle can directly affect genetic medicine. All therapeutic forms of genetic medicine are approved and encouraged by this moral principle. Just as life should be respected, it is also good to promote health, which is the well being of human life. Nontherapeutic genetic experimentation, on the other hand, cannot claim to promote health or to provide therapy; it uses the individual for research purposes. Involuntary nontherapeutic experimentation can be justified only in a utilitarian consequentialist ethical system, as described above. If a prisoner is condemned to death, it would be a social benefit to try out new drugs on that prisoner. This would readily produce greater happiness in a greater number of citizens. But those who maintain an exceptionless moral principle of respecting and promoting human health have no room for nontherapeutic experimentation unless there has been consent and free choice of risking experimentation for the common good.

Procreative Power. The third principle, that of stewardship of the human procreative power, can be divided into the two categories of respect and responsibility.

(1) Respect. The Church has regarded it as evil to destroy directly the

procreative power through contraception and sterilization. This parallels the evil of directly destroying innocent life itself. In fact, Catholic teaching considers both principles as exceptionless moral principles. Some theologians argue for a distinction between the principles. They maintain a virtually exceptionless position on abortion but do find exceptions to the principle of respecting procreative power. Pope John Paul II most recently reiterated the Catholic teaching in his apostolic exhortation *Familiaris Consortio*.[10]

(2) Responsible Use. Responsible use of procreative power clearly has genetic application. One can ask "Would there be times when people with a genetic heritage of disease are acting irresponsibly in having children?" The Pope John Center study, *Genetic Counseling, The Church and The Law*, discusses this difficult question.[11] Does a married couple ever have an *obligation* not to procreate? This is not a question of an obligation to use contraception or sterilization, because there are other means of not procreating. That is a separate question. The Pope John Center study answers the question about not procreating affirmatively in the narrow circumstances where a couple has a 25 percent risk of producing a child with a serious genetic disorder and would be unable to cope with that situation without financial and other assistance.

Application Of Principles

Contemporary progress in genetics suggests some application of these ethical principles. First, in light of the first two principles discussed above, one can certainly regard research and therapeutic medicine as good. Many genetic counselors point out that at least 80 percent of their work has nothing to do with abortion. It helps people understand the source of their genetic problems and what is wrong with their baby. They learn about appropriate treatments, and perhaps understand what future risks they face. Those who are alarmed by the so-called therapeutic abortion of genetically defective infants must not write off the whole field of genetic counseling because of the occasions where it is associated with abortion.

Second, it is appropriate for couples to obtain their genetic history, and there are good reasons for couples to submit voluntarily to screening programs. There are some special problems with compulsory screening because of the potential misuse of such information. But certainly voluntary screening and subsequent morally acceptable steps to avoid conceiving a severely defective child are proper applications of these principles.

An entirely different issue surfaces when a defective human being is already in existence. If genetic diagnosis is performed to prevent the birth of a defective child, it is not accurate to say that genetic medicine has pre-

95

vented the defect. Physicians have prevented the birth but have not really dealt with the defect. They did not in any way provide therapy for the genetic defect. Some of the more outspoken leaders of the right-to-life movement insist that to abort a defective child is neither a therapeutic procedure nor the genuine practice of genetic medicine, but simply exterminative medicine.

In programs on genetic diagnosis, slides are often presented showing the pathetic condition of genetically defective infants or fetuses. Many of those unfortunate babies will not actually face a lifetime of illness and abnormality. Many will die, simply because they are born with such a dramatic genetic defect that they are already dying. But if such a child is not dying, may one take the life of the child? The thought of such killing brings to mind other sets of slides that are equally disturbing, slides that show the horrors of abortion and have been popularized in the right-to-life movement. There is something terribly pathetic in those slides as well.

If one followed the ethical system of intuitionism, one might obey the impulse to supplement nature's work of spontaneous abortion of defective fetuses. But the ethical system of natural moral law and transculturalism teaches respect for even the most defective individual. It recognizes life as a gift and always insists on care for every human being. This system recognizes the limits of prolonging life obligations, but it does not countenance killing defective infants.

A genetic counselor who would suggest that physicians complete God's selection process by aborting defective fetuses attributes to God the causing of genetic defects. I suggest genetic defects are more often caused by environmental conditions and other specific human activities. It may be said that God allows infants or fetuses to be spontaneously aborted for genetic reasons, but this does not offer divine approval to carry out a complete elimination of defective infants. To make that step falsely implies that God kills some defective babies and physicians should kill others. God allows some babies to die, and physicians may find themselves reluctantly watching other babies die, but God does not kill the unborn, and neither should physicians.

Discussion

Q Is it morally acceptable for a Catholic health care facility to provide amniocentesis testing for patients who may then choose abortion?

A If one is reasonably confident that amniocentesis paves the way for a couple to choose an abortion, this raises substantial ethical problems for the facility.

Q But with genetic engineering coming into reality, Catholic facilities

Ethical Principles and Genetic Medicine

should do genetic diagnosis testing in order to give service to people. I am sure Catholic facilities could give good genetic counseling to patients.

A   I am not ready to say that Catholic facilities should never provide genetic counseling and amniocentesis. But I think the special ethical issues involved must be addressed.

First, there are situations, especially in the third trimester of pregnancy, where amniocentesis is used directly to benefit the baby. Such a use is cause for virtually no ethical concern as long as the procedure is done carefully and competently. But, in the great majority of cases where amniocentesis is performed early in the second trimester for diagnostic purposes, there is no direct therapy or benefit for the baby. Indirectly, however, the baby may benefit if the mother becomes calmer and less anxious because of the test results. Hence, one clear ethical problem is to determine whether there is moral justification for submitting an unborn child to the risk of a test that may offer the child practically no benefit when there is no intrauterine therapy for the suspected condition. This problem persists, even though amniocentesis can be offered more safely by competent specialists, and the statistics on risk vary greatly.[12] Should Catholic facilities provide procedures with some degree of risk and no documented therapeutic benefit to the fetus?

The second ethical problem is the facility's cooperation in abortion by helping select those to be aborted. If over 90 percent of women with a positive diagnosis of a genetically defective fetus choose abortion, the agency performing the diagnosis cannot escape cooperation in that outcome, even though it may not approve of or provide abortion.[13] Ideally, a pro-life health care facility should seek an assurance from a woman before testing that she will not seek abortion, since the facility considers abortion as unjust and an immoral practice of prenatal eugenicism. At the very least, the testing agency should offer supportive counseling and all available assistance to mothers with a positive diagnosis of a genetically defective fetus to discourage abortion.

Q   But in offering amniocentesis, an agency is simply offering a test. It does pregnancy testing and has no idea if a woman will then decide to have an abortion.

A   Unfortunately, I think amniocentesis testing for genetic defects raises greater problems than pregnancy testing. If over 90 percent of the women who received pregnancy testing chose an abortion when the test is positive, however, this too would be a problem.

Amniocentesis testing provides specific information on which to base an abortion decision. The testing identifies targets for abortion. It selects innocent individuals to be killed. It is true that other individuals are saved by the selection process. But the process of choosing individuals to be killed resembles the action of an unjust judge who picks out innocent persons for judicial murder to pacify a lynch mob. Society is prepared to "lynch" selected fetuses under suspicion of genetic defect. The judgment rendered by the testing may reduce the number of innocent fetal deaths, but it also involves the testing agency in killing the innocent.

The testing process cannot be adequately justified on the grounds that otherwise a more widespread unjust killing would occur. Referring back to the transcultural exceptionless norm against killing the innocent, this norm does not become less binding on pro-life health care facilities, even if society at large were killing greater numbers of innocent individuals. I thus conclude with my initial comment: amniocentesis for genetic diagnosis raises special ethical problems that must be faced.

Footnotes

1. Germain Gresez, and Joseph M. Boyle, *Life and Death with Liberty and Justice: A Contribution to the Euthanasia Debate.* (Notre Dame, IN: University of Notre Dame Press, 1979).
2. Benedict Ashley and Kevin O'Rourke, *Health Care Ethics, A Theological Analysis.* 2nd ed. (St. Louis: The Catholic Health Association, 1982), pp. 148-175.
3. Alfred Jules Ayer, *Language, Truth and Logic.* (London: Gollancz, 1936).
4. For an implicit intuitionism, see Jean-Jacques Rousseau, *Discourse upon the Origin and Foundation of Inequality Among Mankind in Social Contract and Discourses.* (1753; reprint ed., New York: E. F. Dutton, 1976).
5. Richard Roach, "Consequentialism and the Fifth Commandment," in *Moral Responsibility in Prolonging Life Decisions.* Donald McCarthy and Albert Moraczewski, eds. (St. Louis: Pope John Center, distributor: Franciscan Herald Press, 1981), pp. 20-43.
6. John Stuart Mill, *Utilitarianism and Other Writings.* (1863; reprint ed., Cleveland: New American Library, Meridian Books, 1962).
7. See Vernon J. Bourke, *History of Ethics.* vol. 1 (New York: Doubleday and Co., Image Books, 19709), pp. 147 ff.
8. See Ashley and O'Rourke, pp. 162-175, see also Benedict Ashley, "The Use of Moral Theory by the Church," in *Human Sexuality and Personhood.* (St. Louis: Pope John Center, distributor: Franciscan Herald Press, 1981), pp. 223-242.
9. Sacred Congregation for the Doctrine of the Faith, *Declaration on Euthanasia.* (Washington DC.: USCC Publications Office, 1980), p. 5.
10. Pope John Paul II, *Apostolic Exhortation on the Family.* (Rome: *Familiaris Consortio.* Nov. 22, 1981).

11. *Genetic Counseling, The Church, and The Law.* Gary M. Atkinson and Albert Moraczewski, eds. (St. Louis: Pope John Center, distributor: Franciscan Herald Press, 1980), pp. 123-124.
12. Atkinson and Moraczewski pp. 18-24, 105-107; see also Patricia Monteleone and Albert Moraczewski "Medical and Ethical Aspects of the Prenatal Diagnosis of Genetic Disease," in *New Perspectives on Abortion.* Thomas W. Hilgers. Dennis J. Horan, and David Mall, eds. (Frederick, MD: University Publications of America, Inc., Aletheia Books, 1981) pp. 45-59.
13. See, for example, the 1976 *Facts* publication of the National March of Dimes (White Plaines, NY), which includes a survey of 37 institutions providing amniocentesis among patients at risk of genetic disease. Of the 2,000 women tested, 97 percent carried a normal child. Of the remaining 64 affected fetuses, all but 2 were aborted. See also, *Who Will Defend Michael?* (Export, PA: U.S. Coalition for Life, 1976) p. 8.

chapter 8

Genetic Manipulation
Some Ethical and
Theological Aspects

Rev. Albert S. Moraczewski, OP, PhD

Introduction

In this chapter I will reflect on some of the theological and ethical dimensions of the topics explicitly treated in this book. The technical developments in the area of genetics in general, and in genetic engineering in particular, are truly awesome. They are all the more awesome when one realizes the short time interval over which they took place. In less than 10 years, genetic engineering has moved from fantasy to reality. Granted, the reality reached to date is just the tip of the iceberg, but initial steps already reveal some of the possibilities for genetic manipulation. With that vision of possibilities, one must be concerned about the consequences and the implications of these advances for the human race. Are these technological wonders truly for the benefit of the human race and are not just a bauble of temporary interest or, worse, a release of uncontrollable forces?

Among the statements of scientists who have reflected on these concerns, one strikes me as remarkable. Robert L. Sinsheimer, a biophysicist who is chancellor at the University of California at Santa Cruz, describes the power possibilities of genetics as follows:

> In *Homo sapiens* . . . something new appeared on this small globe. The next step of evolution is ours. We must devise that once again on this sweet planet, a fairer species will arise.[1]

Fr. Moraczewski is vice president for research, Pope John XXIII Medical-Moral Research and Education Center, St. Louis, MO

This quotation implies that the human race once arose as a fairer species as compared to preexisting beings. Society today is on the threshhold of being able to form yet a newer and possibly fairer species. Is this technically possible? And if so, is it an enterprise we should initiate?

Another, perhaps more restrained forecast is expressed by the Nobel Laureate Paul Berg, a microbiologist at Stanford University. He was instrumental in developing such advances and early developed a great concern about the possible dangers of genetic manipulation. To him goes credit for suggesting to his colleagues a moratorium on the use of recombinant deoxyribonucleic acid (DNA) technology.[2] Indeed, that moratorium did take place; government regulations were formulated and subsequently modified several times.[3,4]

Paul Berg has this to say:

> The development and application of recombinant DNA techniques has opened a new era of scientific discovery, one that promises to influence our future in myriad ways. It has already had a dramatic and far reaching impact on the field of genetics, indeed, in all of molecular biology. . . . There is no doubt that the development and application of recombinant DNA techniques has put us on the threshold of new forms of medicine. There are many who contemplate the treatment of crippling genetic diseases through replacement of defective genes by normal counterparts obtained by molecular cloning. Scenarios about how this could be done are rampant, only a few of which are plausible.[5]*

He, as well as many others, have seen recombinant DNA technology as almost a divine blessing, possibly offering a utopia.

At the same time, many in the scientific community wanted to look at the issue more comprehensively. Accordingly, interdisciplinary meetings and conferences were held over several years. Among these were two sponsored by prestigious groups: the National Academy of Sciences and the New York Academy of Sciences.[6,7] Both seriously considered and weighed the societal impact of genetic manipulation and both were serious efforts by scientists and others to ponder responsibly the implications of this new power. Neither—not unexpectedly—received significant theological input.

Accordingly, the two ethical questions I would like to address from a Christian perspective may be stated as follows: We can; *may* we? We can; *must* we?

---

* Copyright © 1981 by the Nobel Foundation, Stockholm, Sweden. Reprinted with permission.

We Can; May We?

The Nature of Human Nature

Just because society has the technology to do something, is the action thereby ethically justified? Once the technology to do something exists, are human beings *obliged* to use it? In answering these kind of questions, a number of value issues are presupposed. What is good? What is the hierarchy of goods? What degree of risk are people willing to take for what kind of benefit? The answers depend on certain values cherished by the society or the individual. Yet not only values, but the world view espoused, are important. What is the ultimate origin of the universe? Where do humans fit in? What are humans? Whence are we and where are we going? What is the purpose of life? All people must grapple with these fundamental questions, and answers are necessary in order to answer specific ethical questions raised about genetic engineering.

What is man (in the generic sense)? Christians can obtain insights from the sacred scriptures which instruct that the universe and humans did not evolve by chance. Rather, in the Judeo-Christian perspective, the universe is the result of a decision and act of an intelligent being, God, of a creation *ex nihilo* done with a purpose and, indeed, with love. Thus, from that perspective the human is necessarily not the highest pinnacle in existence. At the very least there is a being who transcends us and to whom we are responsible. That position contrasts to the prevalent perspective of the culture in which Christians live. Christians eat, drink, and breathe in a culture of secular humanism that places the human at the very top, the pinnacle of reality. Therefore, from that perspective, the human being alone is the measure of himself and of everything else. An additional difference is that Christians believe that humans are redeemed by the Man-God, Jesus Christ.

Those two truths, creation and redemption, will be important for several other questions I will subsequently raise. Hence, we can assert that man is a being created by God and redeemed by Christ for the purpose of participating eternally in the infinite happiness of God. That is a vision to be accepted on faith. This truth cannot be discovered or proved by unaided reason. Although it transcends reason, yet it does not contradict it.

That truth is one aspect of the human that we can learn from revelation, but we can discover other truths and identify other values in revelation. In conjunction with revelation, one should look at what modern science says about the human being insofar as he is observable and measurable. We cannot overlook that contribution. This is perhaps one of the greatest challenges of the time: To relate the results of science in all its myriad forms with the insights of revelation, which gives a more or less broad picture of humankind. Considerable information has been discovered by science about the biological, psychological, and sociological aspects of human beings. Any accurate and comprehensive picture of the human has to consider what scientific research has uncovered as well as what has been received from revelation.

In the recorded histories of centuries past and even before, humans sought help from transcendent forces—God or gods—because they had very little control over nature. More recently, as science and technology rapidly developed, human beings have become more independent of the vagaries of nature. This is as it should be; there is nothing inappropriate in the fact that the human race has grown up, so to speak, and has become increasingly independent of earlier "outside" help without denying its origins, existential dependence, or destiny. Indeed, the creation accounts in Genesis (especially the first two chapters of Genesis) clearly show that God has given human beings a dominion over nature:

> Then God said: "Let us make man in our image, after our likeness. Let them have dominion over the fish of the sea, the birds of the air, and the cattle, and over all the wild animals and all the creatures that crawl on the ground."
>
> God created man in his image;
> in the divine image he created him;
> male and female he created them.
>
> God blessed them, saying: "Be fertile and multiply; fill the earth and subdue it. Have dominion over the fish of the sea, the birds of the air, and all the living things that move on the earth." [Gen. 1:26-28]
>
> The LORD God then took the man and settled him in the garden of Eden, to cultivate and care for it. [Gen. 2:15]
>
> The LORD God said: "It is not good for the man to be alone. I will make a suitable partner for him." So the LORD God formed out of the ground various wild animals and various birds of the air, and he brought them to the man to see what he would call them; whatever the man called each of them would be its name. The man gave names to all the cattle, all the birds of the air, and all the wild animals; but none proved to be the suitable partner for the man. [Gen. 2:18-20]*

---

*Scripture texts used in this chapter are taken from the NEW AMERICAN BIBLE, copyright © 1970, by the Confraternity of Christian Doctrine, Washington, DC, are used by permission of copyright owner. All rights reserved.

Human Dominion Over Nature and Self

That mastery over nature is a delegated dominion: it is limited, and it requires human beings to be accountable to God. The command given to Adam and Eve to be fruitful and multiply, to fill the earth and subdue it, does not mean to destroy it or to master it by brute force as if we were a conquering army, but rather to make intelligent use of it. Responsible stewardship requires that we understand the forces of nature at every level and work with them to better the conditions under which we live so that all can live a more human life.

Unfortunately, we have also discovered that as we progress in our control of nature, we acquire tools of increasing power that can be easily abused. Even our hands, which are our primary instruments, can be abused. These very hands can be used in prayer and to caress as a sign of love, but they can also be used to choke somebody. From the very beginning, tools could be used and misused, as in the story of Cain and Abel (Gen. 4:1-8). Because something can be misused does not *ipso facto* rule out the possibility of appropriate usage. That is important in issues such as genetic engineering. Although genetic manipulation could be misused, such potential misuse does not necessarily rule out appropriate use. The challenge that faces us requires that we seek the wisdom necessary to use properly the powers that we have developed in recent times—a sentiment voiced by Pope John Paul II.[8]

Although human beings have received a delegated dominion over *nature*, it does not necessarily follow we have a dominion over *ourselves*. Clearly, if we did, it would be a limited dominion. What would those limits be? That determination is crucial in the area of genetic engineering, where it is possible to alter the human genome—that collection of genes that makes us what we are.

We cannot take it for granted that we do have dominion over ourselves, even if a delegated dominion. Yet if we did not have any dominion over our bodies we could not use them to eat, drink, walk, and work. Simple common sense gives us the answer. Pope Pius XII stated the matter concisely in the face of a totalitarian state that would dominate its subjects absolutely:

> It is not the community but the Creator Himself Who gives to man dominion over his body and his life, and to the Creator man must answer for the use he make of them.[9]

That last phrase also makes clear the accountability each individual has to God.

Determination of the specific limits of that delegated dominion is not as evident as the fact itself of limited dominion. Urged by the then recently revealed atrocities that took place in the Nazi concentration camps, as well as by the tendency of some to place as supreme the advance of medical knowledge, Pope Pius XII stressed the fact and importance of the limits of human power over the body. Thus, in connection with medical treatment, he stated:

> As far as the patient is concerned, he is not absolute master of himself, of his body, or of his soul. He cannot, therefore, freely dispose of himself as he pleases. Even the motive for which he acts is not by itself either sufficient or determining. The patient is bound by the immanent purposes fixed by nature. He possesses the right to use, limited by natural finality, the faculties and powers of his human nature. Because he is the beneficiary, and not the proprietor, he does not possess unlimited power to allow acts of destruction or of mutilation of anatomic or functional character. But, in virtue of the principle of totality, of his right to employ the services of the organism as a whole, he can give individual parts to destruction or mutilation when and to the extent that it is necessary for the good of his being as a whole, to assure its existence or to avoid, and naturally to repair, grave and lasting damage which could otherwise be neither avoided nor repaired.[10]

Humans have evolved biologically and socially over time, but for how long or what this evolutionary process entails are factors still being worked out. Without deciding which theory of evolutionary process is correct, one can see human evolution taking place as part of God's general providence. Not to be denied is the truth that the spiritual dimension of human beings cannot have evolved from materiality and thus requires a special action of God.[11] One can say, at least, that evidence argues that the human body is the product of a gradual evolutionary process that took place over many hundred thousands or millions of years, depending at what point is considered the beginning.[12] Even if evolution were not directional in the sense that prior to an evolutionary event we could not predict it, or at least we could not have foreseen the development of the human species from the existence of the early mammals, a significant item not to be overlooked is that we are here. From our perspective, the evolutionary development of the universe, of the solar system, and of life on planet earth are under the guidance of divine providence.[13] Although not necessarily admitting (or denying) the existence of a creator, several modern astronomers and physicists have developed an "anthropic principle," which holds that the universe looks *as if* it were designed as a setting for humans.[14,15,16,17] "Why then is the universe the way it is? Because we are here!" is the bold exclamation of one physicist.[18]

Human Control over Evolution

Is further evolution possible? Can the human race as it currently exists evolve further by those natural forces heretofore operative? If so, what kind of change could take place? Would a larger or more complex brain help? Would an additional hand or an arm be an improvement? Would sharper vision, nonfailing hearing, a longer life span, and a better memory be desirable changes? If such changes are possible or desirable, can present technology effect them? As a knowledgeable and deliberate choice, is this a course upon which we want to embark? Who would decide, and how would we go about selecting the decision maker? If we were to redesign the human being, what characteristics should this *Homo futurus* possess? These are some of the problems facing us. Gradually, humankind is acquiring the requisite power. Before the necessary technology to change our bodily structure, and other life forms, becomes fully operative, we might want to reflect on whether we may, whether we must.

We can now alter the genetic composition of bacteria and, presumably and gradually, we will be able to alter the genetic composition of higher animals. We already have seen some benefits from this kind of alteration. Insulin, growth hormone, and interferon can now be made by genetically altered bacteria that would not be able to do so by natural genetic capabilities. By introducing the proper genes into their genome, these bacteria are able to produce these substances for human consumption. Genetic alterations can also take place in agriculture and animal husbandry. These developments can increase both the quantity and quality of the food supply.

New fundamental knowledge, too, is available as a result of genetic research. Chapter 6 outlines how much is now known about the structure of the gene, the sequence of bases in DNA and ribonucleic acid (RNA), and the relationship to the amino acid composition of proteins to their activity. The importance of protein folding, which must take place in a particular way because of the activity of protein as a three-dimensional substance depends on certain components of the protein being accessible to other molecules is now more fully appreciated. Not just folding is significant, since the change of one amino acid can determine whether the protein molecule will fold properly and therefore whether the desired action will occur. Basic knowledge about genes and their bearing on protein structures and activity has been the result of research to find answers to practical problems.

Possible Risks

The improved control we have recently obtained over living organisms through genetic manipulation has not been, and is not yet, without some continuing risks. Not without some basis, concern has been expressed that by the deliberate decision and action of individuals, or terrorists groups, genetically altered organisms would be introduced into our environment and would be "superlethal" for humans. The human body's ordinary defense mechanisms would not be able to cope with them because the organism would have totally new powers. A similar fear motivated the National Aeronautics and Space Administration (NASA) to quarantine the first few astronauts who had ventured on the moon's surface. When it was discovered that neither they nor their vehicles harbored alien pathogens, the practice was discontinued.

Deliberate attempts to destroy human life have not been the only concern. Fear has been expressed, too, that pathogenic organisms might be produced inadvertently, against which antibiotics or other natural defenses would be totally ineffective. Indeed, as noted above one of the early workers in the field of recombinant DNA, Dr. Paul Berg, urged his colleagues to observe a moratorium on further research in that technology until there had been sufficient time to assess what was already known about the organisms with a newly altered genome. Subsequently, the relevant agencies of the United States government issued strict guidelines for research and development in this field. As more experience was gained and the potential hazards more clearly recognized and evaluated, these guidelines have been progressively relaxed but not discarded.

One hazard of special concern has been that some food supplies are particularly vulnerable because, for example, much corn and wheat are each of one, or closely related strains, so that a pathogen for corn could destroy most of the crop in an entire region, and similarly for wheat. Agriculturalists have become sensitive to this danger and are striving to maintain genetic variability among various food plants as well as to build up a reserve stock in safe storage of seeds, i.e., with a wide spectrum of responses and growth patterns.

Although oil is an essential energy source, spillage from ruptured pipes and shipwrecks has caused considerable damage to aquatic life and beaches. The possibility of developing and employing "oil-eating" bacteria was considered and successfully carried out. It was feared, however, that should such bacteria escape into the environment, they could "eat" all the oil in cars, machinery, and oil wells. This would be a catastrophe.

Consequently, in assessing whether we should alter the genetic composition of infrahuman life, we must consider the possible benefits as well as reasonably foreseen risks. Because of the nature of the forces that might be unleashed, wisdom recommends that we move slowly until we know more.

The Future Universe and Humans

What is the precise relationship of the human race to the universe? We are part of it, indubitably, and our bodies are made up of the same stuff. Yet our spiritual nature transcends the material universe in all its mysterious vastness. When God instructed Adam and Eve to "be fertile and multiply; fill the earth and subdue it" (Gen. 1:28), what kind of dominion did he grant? Clearly it involved responsibility and accountability. But did he mean we should preserve unchanged what is here or shape and alter the "raw" material according to our needs and desires?

Many Christians have changed their attitude on the relationship of the present to the eternal future. Until recent times, many, and perhaps even the majority, thought that the earth was a huge waiting room where we were all quietly seated, marking time, before entering heaven. Another attitude has become increasingly visible. There is no waiting room, just a half-built house. On earth we have a creation to complete. We have a task to perform; we are not just waiting. What we do today affects the future. The end of time is partly a function of what we do with the present. On a cosmic scale, we are not merely passive observers, but we are—within divine providence—active participants responsible for shaping, directing, and influencing the world and the time frame of events.

The end of this world, as we now imagine it, is not an arbitrary cutoff point determined by God but apparently is conditional. Just as his dealings with the Israelites of the Old Testament were conditional upon their fidelity to the covenant, so too the continued history of the human race, yet to be written, is conditioned on our accomplishing certain tasks. Those tasks primarily are directed to establishing God's kingdom—an order of justice, love, and peace. This effort is not merely a human endeavor, however, but the result of the Holy Spirit's action in us to achieve his purpose. Human effort nonetheless is required of us in a fashion similar to that required of the Israelites when they conquered and possessed the Promised Land.

One aspect of the development of the kingdom is human progress, which includes advances in technology. The appropriate use of human discovery and invention is necessary in order to provide for the growing number of human beings. In addition, the detailed understanding of the universe (and its component parts) that science can furnish, and the control over matter and material forces that technology can exercise, allow increasing numbers of humans to be sufficiently liberated from the restrictions and adversities of the sin-wounded environment to have the opportunity to know, love, and serve God and one another. Thus, printing, electricity, electronic communication devices, transportation, and distribution have enabled religious and secular knowledge to spread over the entire surface of the earth.

Vatican Council II addressed itself to this issue in the following words:

33. Man has always striven to develop his life through his mind and his work; today his efforts have achieved a measure of success, for he has extended and continues to extend his mastery over nearly all spheres of nature thanks to science and technology. Thanks above all to an increase in all kinds of interchange between nations the human family is gradually coming to recognize itself and constitute itself as one single community over the whole earth. As a result man now produces by his own enterprise many things which in former times he looked for from heavenly powers.

In the face of this immense enterprise now involving the whole human race men are troubled by many questionings. What is the meaning and value of this feverish activity? How ought all of these things be used? To what goal is all this individual and collective enterprise heading? The Church is guardian of the heritage of the divine Word and draws religious and moral principles from it, but she does not always have a ready answer to every question. Still, she is eager to associate the light of revelation with the experience of mankind in trying to clarify the course upon which mankind has just entered.

34. Individual and collective activity, that monumental effort of man through the centuries to improve the circumstances of the world, presents no problem to believers: considered in itself, it corresponds to the plan of God. Man was created in God's image and was commanded to conquer the earth with all it contains and to rule the world in justice and holiness: he was to acknowledge God as maker of all things and relate himself and the totality of creation to him, so that through the dominion of all things by man the name of God would be majestic in all the earth.

This holds good also for our daily work. When men and women provide for themselves and their families in such a way as to be of service to their community as well, they can rightly look upon their work as a prolongation of the work of the creator, a service to their fellow men, and their personal contribution to the fulfillment in history of the divine plan.

Far from considering the conquests of man's genius and courage as opposed to God's power as if he set himself up as a rival to the creator, Christians ought to be convinced that the achievements of the human race are a sign of God's greatness and the fulfillment of his mysterious design.

With an increase in human power comes a broadening of responsibility on the part of individuals and communities: there is no question, then, of the Christian message inhibiting men from building up the world or making them disinterested in the good of their fellows: on the contrary it is an incentive to do these very things.[19]

Clearly, technological advances are a good and should not be simply dismissed by Christians as evil, even if in their application misuse may ensue occasionally.

Technology and the Good of Humans

Not to be overlooked is the contemplative side of science and technology. Inventions such as microscopes, telescopes, and space vehicles which have sent back information to earth about the various planets and moons, have revealed aspects of the universe that otherwise would have been hidden. The scientific study of subatomic structures, of the biochemistry of living cells, and of the embryological development of the human body all have revealed awesome design and order. Indeed, wherever and whatever in this world that science has probed, analyzed, and pondered has shown incredible complexity and beauty to tantalize the imagination and to stimulate the mind. For the believer, the revelation of God's mystery in creation leads to the prayer of praise.

Yet this strong affirmation of science and technology does not mean that there is no danger or cause for concern. All too readily human hubris has led to a misuse of technology. In modern times the discovery and application of nuclear power has opened the possibility of a nuclear holocaust. Genetic engineering has given humans another power, also potentially destructive of life as we know it. That power could radically alter existing life forms to produce entirely new species. Not wanting to be silent on the matter, the U. S. bishops issued a statement praising the benefits but also expressing deep concern about the hazards associated with genetic engineering. They also provided some moral guidelines in this new field of human activity.[20]

Ultimately, it becomes necessary to ask how can we determine what use of technology is appropriate and good for humans. Pope John Paul II began to address that issue in his first encyclical, *Redemptor Hominis:*

The development of technology and the development of contemporary civilization, which is marked by the ascendancy of technology, demand a porportional development of morals and ethics. For the present, this last development seems unfortunately to be always left behind. Accordingly, in spite of the marvel of this progress, in which it is difficult not to see also authentic signs of man's greatness, signs that in their creative seeds were revealed to us in the pages of the Book of Genesis, as early as where it describes man's creation, this progress cannot fail to give rise to disquiet on many counts. The first reason for disquiet concerns the essential and fundamental question: Does this progress, which has man for its author and promoter, make human life on earth "more human" in every aspect of that life? Does it make it more "worthy of man"? There can be no doubt that in various aspects it does. But the question keeps coming back with regard to what is most essential—whether in the context of this progress man, as man, is becoming truly better, that is to say more mature spiritually, more aware of the dignity of this humanity, more responsible, more open to others, especially the neediest and the weakest, and readier to give and to aid all.[21]

Technology to be in the true service of humans, must not, at the very least, prevent them from reaching eternal life and, as a rule, should assist each person to become more truly human, more truly free from external and internal shackles, and more likely to achieve the goal of union with God. Thus, genetic alterations of infrahuman life may be done if the new life form is judged to be advantageous for a fuller human life, e.g., food production could be improved, diseases better controlled, healing more complete and rapid, and the environment generally made more healthful and pleasant.

May We Redesign Human Nature?

Above, I briefly discussed our technical ability to modify the genetic makeup of a human, initially with the intent of correcting certain defects. Eventually, it may be possible to change radically the genome of individuals and their progeny. But *may* we? What principles do we use to answer that question? One of the nonnegotiable givens from a Christian point of view is that our goal is union with God. Therefore, anything that would interfere with that union is considered inimical to human good and hence morally bad. Contrariwise, anything that would help relating to God and being united to him would ultimately be considered good. Hence, no genetic alteration that would impede that union could be acceptable. Such an alteration could, for example, make the individual radically incapable of making a morally responsible decision.

But what would constitute a change of human nature? When would a new species be formed? One test of a species difference is that individuals from the two populations are not able to generate fertile offspring. But the converse is not necessarily true. Infertility between two individuals can be caused by factors other than species difference. Furthermore, individuals from different species may not produce offspring because of behavioral differences and not because of genetic incompatibility. The current human population is all of one species, since every human race is apparently cross-fertile with every other race as far as available evidence indicates.

Genetically adding another finger, arm, or leg would not seem to result in a radical enough change to constitute another species. But to alter the human body in a manner so that such individuals would think in a radically different manner, or so that the sense organs were absent or sensitive to a different array of stimuli that would provide a different perception of the world, probably would result in individuals of a different species. Such individuals would experience the world in a very different manner, so that communication and community would be highly improbable or even impossible. Hence, from this perspective, individuals so altered would not be members of the human species.

If the individual lacked radically the capacity of free will, certainly that would not be human. This condition should be clearly distinguished from those humans who have the radical capacity for free will but are not capable of exercising it because of some defect, disorder, or accident.

Another consideration is relevant here. C. S. Lewis, the former Anglican scholar of Medieval and Renaissance Literature at Cambridge University, makes an interesting point in *Perelandra,* the second volume of his theological-science fiction trilogy.[22] The action takes place primarily on the planet Venus, where a new race of rational animals has begun. Except for the green color of the first male and female, they are of human shape, size, and behavior—which, in the book, is a behavior before sin. The author observes through one of the characters that since the Son of God has taken on, in a personal union, our human nature, including, of course, the body, no other bodily form would do for a *rational* being.

Consequently, one could argue that human nature, having been so dignified and elevated by the Son of God, would be dishonored if a significant alteration were made. This position needs further reflection and development before many, I believe, would find it persuasive.

We Can; Must We?

Do we have an obligation to develop technological capabilities in the field of genetic manipulation? As a point of departure, Paul Berg's words are relevant:

> The advent and widespread use of the recombinant DNA technology for basic and medical research and the implications for industrial and pharmaceutical application has also revealed, or perhaps created, an underlying apprehension, an apprehension about probing the nature of life itself, a questioning of whether certain inquiries at the edge of our knowledge and our ignorance should cease for fear of what we could discover or create. I prefer the more optimistic and uplifting view expressed by Sir Peter Medawar in his essay entitled "On the Effecting of All Things Possible."
>
> "If we imagine the evolution of living organisms compressed into a year of cosmic time, then the evolution of man has occupied a day. Only during the past 10 to 15 minutes of the human day has our life been anything but precarious. We are still beginners and may hope to improve. To deride the hope of progress is the ultimate fatuity, the last word in the poverty of spirit and meanness of mind . . .
>
> This passage speaks of the need to proceed. The recombinant DNA breakthrough has provided us with a new and powerful approach to the questions that have intrigued and plagued man for centuries. I, for one, would not shrink from that challenge.[23]

If our understanding of the human race's development is anywhere near the truth, then we must recognize that unless our primitive ancestors had pushed forward and developed a technology, of which stone knives, axes, and fire may have been the early examples, the likelihood that we would have survived except as few, isolated individuals scattered over a relatively small proportion of the earth's surface is indeed very slim. Clearly, there was a need and an imperative—fill the earth and subdue it—to develop a technology and alter the face of the earth: cultivate plants, hunt and herd animals, process hides and plant fibers for clothing, cut trees, build shelters, bridge streams, and cut and carve stones, bones, and wood tools, decoration, and communication.

Once the process of technological development began, the rate gradually increased. As the human race grew, the land could no longer support the

---

Copyright © 1981 by the Nobel Foundation, Stockholm, Sweden. Reprinted with permission.

population, and dispersal was inevitable. Improved means of transportation and communication appeared. A new need arose, and a new technology was developed.

Coupled with the pressure of meeting the basic requirements of food, clothing, and shelter was a tremendous inner drive to understand, to explore the unknown, and to master and overcome the obstacles of nature. Disease, injury, and death were constant companions to be feared, controlled, and conquered. Human ingenuity responded, sometimes slowly and sometimes quickly, but ultimately with some success.

In retrospect, these multitudinous advances of science and technology were necessary in their assembly if the human race was to survive and prosper. Clearly, too, the Church has recognized the essential contributions of science and technology to human welfare, as reflected in the words of Pope John XXIII:

> For it is indeed clear that the Church always taught and continues to teach that advances in science and technology and the prosperity resulting therefrom, are truly to be counted as good things and regarded as signs of the progress of civilization. But the Church likewise teaches that goods of this kind are to be judged properly in accordance with their natures: they are always to be considered as instruments for man's use, the better to achieve his highest end: that he can then the more easily improve himself, and in both the natural and supernatural orders.[24]

> But the progress of science and the inventions of technology show above all the infinite greatness of God, Who created the universe and man himself.[25]

Vatican II also has a few words on the topic:

> Man has always striven to develop his life through his mind and his work; today his efforts have achieved a measure of success, for he has extended and continues to extend his mastery over nearly all spheres of nature thanks to science and technology.[26]

> Far from considering the conquests of man's genious and courage as opposed to God's power as if he set himself up as a rival to the creator, Christians ought to be convinced that the achievements of the human race are a sign of God's greatness and the fulfillment of his mysterious design.[27]

But is it possible that the human race, having developed so far, no longer has a mandate to develop new technologies, especially in light of their highly destructive potentialities, e.g., nuclear bombs, toxic nerve gases, and genetically altered "superbeings"? Discoveries in medicine have helped to "conquer" many infectious diseases; advances in agriculture have increased food production, while transportation and communication technologies have shrunk the world so that a single human community becomes technically feasible even if human frailty and malice greatly impede its realization. Space exploration has enlarged and enriched our vision of the universe, of which our planet is a part.

Adorned with an array of success and failures, the human race is still challenged to subdue the earth until God's kingdom is full realized. In the meanwhile, a balance must be struck between the ruthless use of technological power and the inertness of a nostalgic antitechnology. That balance can be achieved by a bold bonding of science and ethics, as Pope John Paul II urges:

> The essential meaning of this "kingship" and "dominion" of man over the visible world, which the Creator himself gave man for his task, consists in the priority of ethics over technology, in the primacy of the person over things, and in the superiority of spirit over matter.[28]

> There is no doubt that from many points of view technical progress born of scientific discoveries, helps man to solve very serious problems, such as food, energy, the struggle against certain diseases more than ever widespread in the third world countries . . . In order to prevent science and technique from becoming slaves to the will for power of tyrannical forces . . . what is necessary . . . is . . . faithfulness to the moral norms that regulate man's life. It is incumbent on scientists of different disciplines . . . to use all your prestige in order that scientific implications abide by moral norms in view of the protection and development of human life.[29]

Without a balance between the technological and the ethical, the former can develop uncontrollable powers and overwhelm this world. Such can happen not because of any intrinsic evil found in technology (or technologists), but because of the pervasive influence of original sin. Every human discovery, no matter how exalted or noble, can be distorted and have evil effects. The march towards the eschaton, which began when we received our eviction notice from the garden of Eden, has been slowed down much as was the Israelites' march to the Promised Land by their 40-year delay in the waste land of Sinai because of their infidelity to the covenant.

In conclusion, not only may we but there seems to be an imperative to continue the development and application of technology. Such is necessary to make the world an increasingly more suitable place to live for the ever-greater number of human beings. Yet present and future technological endeavors in the field of genetics—or indeed technology generally—must be directed and tempered by an ethic inspired by the wisdom of the Gospel.

Conclusion

The above conclusion is beautifully supported by a recent statement of Pope John Paul II, on October 23, 1982, made to those participating in a study week on biological research sponsored by the Pontifical Academy of Science. After having linked the importance of scientific knowledge of corporeal reality for the life of the spirit, Pope John Paul II stated:

> Consequently I have no reason to be apprehensive for those experiments in biology that are performed by scientists who, like you, have a profound respect for the human person, since I am sure they will contribute to the integral well-being of man.[30]

The Pope then proceeds to expand a little on the scientific research about which he is commenting:

> I have learned with satisfaction that among the themes discussed during your study week you have focused attention on in vitro experiments which have yielded results in the care of diseases related to chromosome defects.
>
> It is also to be hoped, with reference to your activities, that the new techniques of modification of the genetic code, in particular cases of genetic or chromosomic diseases, will be a motive of hope for the great number of people affected by those maladies.
>
> It can also be thought that, through the transfer of genes, certain specific diseases can be cured, such as sickle-cell anemia, which in many countries affects individuals of the same ethnic origin. It should likewise be recalled that some hereditary diseases can be avoided through progress in biological experimentation.
>
> The research of modern biology gives hope that the transfer and mutations of genes can ameliorate the condition of those who are affected by chromosomic diseases; in this way the smallest and

weakest of human beings can be cured during their intrauterine life or in the period immediately after birth.

Finally, I wish to recall, along with the few cases which I have cited that benefit from biological experimentation, the important advantages that come from the increase of food products and from the formation of new vegetal species for the benefit of all, especially people most in need.[31]

Then, to be sure no one underestimates the importance of the moral dimension in scientific research, Pope John Paul II adds:

> In terminating these reflections of mine, which show how much I approve and support your worthy researches, I reaffirm that they must all be subject to moral principles and values which respect and realize in its fullness the dignity of man.[32]

These words of Pope John Paul II both of support and caution for scientific research and technological development do not stand alone. They represent a teaching which he stated in his first encyclical, *Redemptor Hominis,* and has repeated several times since. Consequently, Catholics who have accepted Pope John Paul II's teaching on this matter can proceed with hope and joy to work and support research in the field of genetic engineering.

Footnotes

1. Robert L. Sinsheimer, quoted in Vance Packard, *The People Shapers* (Boston: Little, Brown & Co., 1977), p. 230.
2. Ray Curtiss, III, "Genetic Manipulation of Microorganisms: Potential Benefits and Biohazards," *Annual Review of Microbiology* 30 (1976): 511.
3. Guidelines for Research Involving Recombinant DNA Molecules, Bethesda, MD; National Institutes of Health, June 23, 1976.
4. *Federal Register,* Jan. 29, 1980, 45:6724; July 1, 1981, 46:34462; April 21, 1982, 47:17180.
5. Paul Berg, "Dissections and Reconstructions of Genes and Chromosomes" *Science* 213 (1981):302.
6. National Academy of Sciences, *Research with Recombinant DNA,* (Washington, DC: NAS, 1977).
7. Marc Lappé and Robert S. Morison, editors, "Ethical and Scientific Issues Posed by Human Uses of Molecular Genetics," *Annals of the New York Academy of Sciences* 265(1976):1-208.
8. Pope John Paul II, address to participants of a study week sponsored by The Pontifical Academy of Sciences, Oct. 23, 1982, English trans. *Origins,* Nov. 4, 1982, pp 342-343.
9. Pope Pius XII, Allocution to the Italian Medical-Biological Union of St. Luke, November 12, 1944, English trans. in *The Human Body* (Boston: The Daugh-

ters of St. Paul, 1960), no. 51.
10. Pope Pius XII, Allocution to the First International Congress of Histopathology of the Nervous System, September 13, 1952, English trans. in *The Human Body* (Boston: The Daughters of St. Paul, 1960), no. 359.
11. Pope Pius XII, *Humani Generis* no. 36., Aug. 12, 1950; English trans., National Catholic Welfare Conference.
12. See Sir John Eccles, *The Human Mystery,* (Berlin: Springer-Verlag 1979), pp 74-97.
13. Albert Moraczewski, "A Vision of the Universe in the Spirit of St. Albert the Great," *Spirituality Today* (1980): 305-313.
14. R. H. Dicke, "Dirac's Cosmology and Mach's Principle," *Nature* 192 (1961): 440-441.
15. B. Carter, "Large Number Coincidences and the Anthropic Principle in Cosmology," in Proceedings of Extraordinary General Assembly of International Astronomical Union (Krakov), ed. M. S. Longair (Boston: Reidel, 1974), cited by Eccles, *op. cit,* p. 30.
16. J. A. Wheeler, "The Universe as Home for Man," *American Scientist* (1974): 683-691.
17. George Gole, "The Anthropic Principle," *Scientific American* 245 (1981): 154-171.
18. See Virginia Trimble, "Cosmology: Man's Place in the Universe," *American Scientist* 65 (1977): 85.
19. Vatican Council II, Pastoral Constitution on the Church in the Modern World, nos. 33 and 34, English trans. Austin Flannery, ed. 1975, pp. 933-934.
20. National Conference of Catholic Bishops, "Statement on Recombinant DNA Research," May 2, 1977; see also Bp. Thomas Kelly, Claire Rondall, and Rabbi Bernard Mondelbaum, "Statement on the Control of New Life Forms," *Origins* 10 (1980):98-99.
21. Pope John Paul II, *Redemptor Hominis,* March 4, 1979, English trans. (Washington, DC: United States Catholic Conference, 1979), pp. 47-48.
22. C. S. Lewis, *Perelandra,* (New York: MacMillan, 1944).
23. Paul Berg, "Dissection and Reconstructions of Genes and Chromosomes," *Science* 213 (1981):302.
24. Pope John XXIII, *Mater et Magistra,* no. 246, May 15, 1961.
25. Pope John XXIII, *Pacem in Terris,* no. 3, April 11, 1963, trans. National Catholic Welfare Conference.
26. Vatican Council II, *Gaudium et Spes,* no. 33, December 7, 1965.
27. *Gaudium et Spes,* no. 34, trans., Vatican Council II, Austin Flannery, OP, ed.
28. Pope John Paul II, *Redemptor Hominis,* no. 16, March 4, 1979.
29. Pope John Paul II, *L'Obsservatore Romano,* English ed., April 9, 1979, p. 5.
30. Pope John Paul II, *Origins,* Nov. 4, 1982, p. 342.
31. Pope John Paul II, *Origins,* Nov. 4, 1982, p. 342.
32. Pope John Paul II, *Origins,* Nov. 4, 1982, p. 343.

chapter 9

Legal Issues Relating to Genetic Diagnostic Procedures

by G. Edward Fitzgerald, LLB

Cases involving genetic diagnostic procedures will undoubtedly come before the courts more frequently as medical technology in the field of genetics continues to expand. This chapter will present some recent court decisions on genetic issues and discuss the current state of legal issues surrounding cases involving genetic diagnostic procedures.

California: The *Curlender* Case

California has come to the forefront of genetic disease cases because of the *Curlender* case, in which it was recognized by the Appellate Court that not only do parents have the right to sue for problems relating from the birth of a defective child, but the child can also sue.[1] For purposes of discussion, I would like to divide the *Curlender* case into two parts: Curlender 1 and Curlender 2. Curlender 1 concerns a hypothetical set of facts, i.e., nothing was proved and the court made a decision based entirely on the assumption that certain facts were true. The Curlenders alleged in their complaint that one physician and two medical laboratories gave them erroneous or negligent advice, thereby resulting in their having a child with Tay-Sachs disease. The child, of course, will die sometime between the fourth and fifth year of life. The defendants simply answered the complaint and denied the allegation. The parents then brought another action based on behalf of the child, stating that the child also had a cause of action for damages sustained. The

---

Mr. Fitzgerald is with Gibson, Dunn & Crutcher in Los Angeles.

defendants filed a *demurrer,* which is a motion that acknowledges the truth of the complaint but says that no cause of action is possible.

No state, until June, 1980, had said that a child could sue, although a number of states have evaluated whether parents may sue in those circumstances. Faced with this set of facts, the trial judge determined that there was no reason for him to change the state of law, and he stated that for fundamental policy reasons, the child could not sue.

The case then went to the next level, the three-judge Court of Appeals, in which the Court wrote the opinion that for every wrong there must be a right and determined that the child did in fact have a cause of action. The Court determined that if the physicians had been negligent in performing or authorizing genetic tests and if the laboratories had provided erroneous test results, the child could sue. A great deal of what the judges relied on was not prior law. In general, judges first look at past decisions. They determine what decisions are persuasive, what line of reasoning they like, and what particular policy argument appeals to them, and based upon those factors they forge a decision for the particular set of facts before them.

The court decided that the child had a cause of action that was independent of the parents'. They relied quite heavily on the language of the 1973 United States Supreme Court in *Roe v. Wade,* which indicated to them that if the court had jurisdiction over such a fundamental issue as abortion, the court also had jurisdiction over an existing child.[2] They seemed simply to assume, without any analysis, that if there is a constitutionally protected right to terminate existence, there must also be a right to sue if that existence is defective. In giving the child this cause of action, they went on to say that the child might very well have a cause of action not only against the physician, medical profession, laboratories, health care facilities, and others potentially involved in the transmitting of genetic information, but also against the parents themselves.

But how is a defective child to seek redress and sue his or her own parents? Who will do this? Does the court or a relative appoint a guardian? How does the lawsuit proceed? Moreover, if the parents were reasonably sure that their child would be genetically deformed, how do they and the physician protect themselves legally? Do they exercise their right under *Roe v. Wade* to terminate existence and inexorably eliminate the lawsuit? These are questions that arise from giving a child a separate cause of action. The plaintiff's attorneys who were seeking to give the Curlender child a cause of action were really, I think, attempting not so much to forge new law but to obtain a new remedy that would allow a multiple recovery.

The law in California was already fairly clear. If a physician, hospital, or laboratory was guilty of malpractice in performing a genetic test, the parents could sue. The plaintiff's attorneys emphasized, for their own purposes, the New Jersey case in *Quinlan,* stating that if the courts were willing to decide when life may be terminated, the court should be able to determine that death was preferable to life in a particular instance.[3] It may not be logical, but there is usually no statute to cover these particular problems, and courts must therefore decide them based on their individual evaluation of the policy questions that are presented in the gentic disease cases. This court, in Curlender 1, found that under the facts as presented to it, there was an injury recognized by the law: the birth of a child with a deficit. It stated that the certainty of genetic impairment is no longer a mystery, and this indicated that within certain limitations, such as taking into account the life expectancy of a child with Tay-Sachs disease, the defendants would be liable for the consequences of that injury. At this stage of the proceeding in California, the defendants were then faced with going to court with two cases against them. One of the defendants determined, for economic reasons, to settle the cases and pay Mr. and Mrs. Curlender a sum of money, thus achieving a dismissal of both actions. Nothing was paid to the child.

That didn't end the Curlender litigation, however, because under California law, the settling defendant has the right to pursue a cause of action against the other two parties remaining in this case, the physician and the other laboratory. I expect this second case to be tried next year, and it will be decided not on hypothetical facts, but on real facts.

Mr. and Mrs. Curlender were middle-class people of above-average intelligence and Jewish origin. They had been married for some time. They decided in late 1976 to have a child. Mrs. Curlender had read of a genetic disease involving Ashkenazic Jews, i.e., Jews of Eastern European or European origin, that might affect her. She was concerned about that. Her husband, Hyam, was a Sephardic Jew and thus was not likely to be a carrier of Tay-Sachs disease. One of 30 people with Ashkenazic background will be a carrier; the statistics dealing with the population at large are considerably different.

Mrs. Curlender called her family physician and told him she would like to be tested for Tay-Sachs disease. The physician was not an expert in genetic disease. He recommended a community screening program. Since that was not convenient for Mrs. Curlender, the physician determined that screening could be done by one of two laboratories.

Testimony then gets a little confused. Mrs. Curlender says that on one Saturday, early in 1977, she went to Laboratory A and had her blood test taken. A couple of days later she called her physician. He told her it would be safe for her to have children, and on the basis of his statement, the couple conceived a child. The child had Tay-Sachs disease. Laboratory A apparently sent two specimens to Laboratory B; all A did was take the blood and reduce it to serum. The test is done from the serum. Laboratory A sent two specimens to Laboratory B, one labeled Mr. Curlender and one labeled Mrs. Curlender. Laboratory B performed two tests and returned the results together with a range of normalities that indicated one was in the carrier range and one was in the low end of the normal range. In addition, they sent back a reference to certain articles dealing with Tay-Sachs disease. The articles indicated that couples who were both carriers, even if one was borderline, should consider undergoing additional tests. Thus, the issues that will ultimately be presented when Curlender 2 is tried are whether the physician should have engaged in genetic counseling, and whether he was under any obligation to advise the Curlenders that one of them was definitely a carrier despite the fact that two carriers are needed to produce a defective child. If he had advised both parents of the two results and if the testimony is correct that only Mrs. Curlender was tested, the Curlenders should have spoken up and said that only Mrs. Curlender had a test. These are the primary issues that will be presented concerning the physician. As to Laboratory A, the question to be presented is much more mundane. Did the laboratory act properly in taking the blood specimen? Did it mislabel specimens? If Mr. Curlender was indeed not tested, how did the laboratory label a specimen as his? Laboratory B, the one that chose to settle the lawsuit, is in a much different position. It actually did the test. It used a range of normalities that some experts considered to be outdated. In addition, there was some question whether they should do this test when screening tests were given free, in the Los Angeles community.

Turpin v. Sortini

Another recent California case, *Turpin v. Sortini,* involved an infant afflicted with hereditary deafness.[4] It was a strange case in that the couple already had one child with the disease but had allegedly been advised by their physician that this second child's hearing would be normal and that it was safe to proceed with the pregnancy. According to the couple, however, their first child was totally deaf, and they stated in their complaint that they would not have had the second child if they had known the child would be deaf.

When *Curlender* 1 was decided on a hypothetical basis, the California Supreme Court denied a hearing. In the *Turpin* case, by a 2 to 1 majority, the Court of Appeal disagreed with the reasoning in *Curlender* and stated that the child did not have a cause of action. The California Supreme Court has granted a hearing on *Turpin,* ostensibly to correct, rewrite, or affirm the decision. Normally, the Supreme Court does not grant a hearing simply to affirm a decision; I assume the new decision will bring *Turpin* more in line with *Curlender* and give the child some type of cause of action. I hope the court also questions whether the child should have a cause of action against its parents, because I think that creates some of the more serious problems in litigation on genetic testing.

New Jersey Cases

California, however, is not the only state that has had difficulty in determining the issue of genetic testing. New Jersey, for example, in 1967 decided *Gleitman v. Cosgrove*, which was then the main genetics case in the United States.[5] The Gleitmans had brought a malpractice action against their physician because their child was born with seriously impaired sight, speech, and hearing. Mrs. Gleitman had contracted rubella during the first trimester of pregnancy. She had allegedly been assured by her physician that this would cause her no harm. She also alleged that such an opinion deviated from the state of medical knowledge at that time. The majority of that court denied recovery on behalf of either the parents or the child, stating that they could not compute the damages in the case and it was against public policy to enter into the question of life or death. The court emphasized the difficulty it would have in measuring the difference between life with defects versus the void of nonexistence and determined that it was simply not able to cope with those issues (see Chapter 7).

Even in 1967 in New Jersey there was no legal consensus on this issue. A minority opinion in *Gleitman* stated that it was incompatible with the standard of American court law to permit a wrong, with serious consequential injuries, to go wholly unredressed.

Twelve years later, the issue again came to the New Jersey Supreme Court in *Berman v. Allen*, which involved Mrs. Berman, a pregnant woman in her late 30s who had a substantial risk of having a child with Down syndrome.[6] The child was, in fact, born with that disease. The issue in the case revolved around the allegation that the physician never told the Bermans about amniocentesis and that they should have had additional information on which to base a decision about terminating the pregnancy.

The New Jersey Supreme Court determined that in *Berman*, the parents could sue. They eliminated the cause of action for the child based on their analysis that life is more precious than death, but as far as the parents were concerned, there was no public policy to prevent them from suing for their own damages, i.e., the additional cost of raising a defective child, or for the emotional distress involved in raising such a child.

New York and Pennsylvania Cases

New York also had a difficult series of cases, which are virtually impossible to reconcile. *Park v. Chessin*, in 1977, involved plaintiffs with a child born with polycystic kidney disease, a fatal hereditary ailment.[7] The parents consulted their physician and were allegedly assured that the condition would never reoccur in later births. They had another child; the child had the disease; and the court held that the parents and the child had a cause of action because decisional law must keep pace with expanding technological, economic, and social change. Inherent in this analysis was the recognition that the parents had a statutory right not to have the child based on the abolition of the statutory ban on abortion.

In 1978 the New York appellate courts were faced with two cases concerning mongoloid infants.[8] The parents had allegedly not been told about amniocentesis; the court determined that the parents in each case could recover, but not the children. (The two cases, *Park* and *Becker*, were consolidated for appeal.) The court specifically rejected the idea that a child may expect life without deformity.

In 1977, another New York court, in the Tay-Sachs case of *Howard v. Lecher*, indicated that neither the parent nor the child could recover, although the child's case by that time was deemed moot because the child had already died.[9] The majority decision stated that the parents' cause of action would require the existence of traditional tort concepts beyond manageable grounds. The dissent simply indicated that the case was another malpractice case, i.e., whether the physician acted within the standards of practice.

In 1979, the Pennyslvania Supreme Court again recognized the parents' right, but not the infant's in *Speck v. Finegold*, which involved a malpractice suit by the parents of a child with neurofibromatosis, a serious crippling condition that was already evident in their older children.[10] In a federal district court case in Pennsylvania involving Tay-Sachs disease *(Gildiner v. Thomas Jefferson University Hospital)*, the parents were tested for Tay-Sachs disease and were later determined to require amniocentesis, which was allegedly performed negligently. In any event, the couple had a child with Tay-Sachs disease. The court held that the parents could recover, but not the child.

The Basis of Decisions

These decisions on genetic diagnostic procedures are based on where a court stands in parent-child cases, malpractice, and product liability and its attitude towards tort cases in general. The courts are assuming that genetic cases must be handled by reference to prior case law. For example, almost all the cases discussed in this chapter subscribe to the general philosophy that a wrong can always be remedied. Many courts analyze old cases dealing with illegitimacy and spend a great deal of time coping with these decisions. For example, in California illegitimate children cannot sue their parents because of their birth. The judge in *Curlender* had a great deal of difficulty with that concept. He ultimately decided that there is somehow a difference between being born with a birth defect and being born out of wedlock and that by now the social stigma attached to illegitimacy is not sufficient to warrant changing the law.

What is policy to one court may not be policy to another. Nevertheless, the courts must contend with certain fundamental issues: Should a child, born with a defective condition that presumably could have been determined before birth, have the right to sue his or her own parents? Should the parents have the right to raise their religious beliefs as a defense? How about First Amendment rights? No court has yet considered them, but giving children a cause of action against their parents inevitably raises First Amendment questions.

But how do we provide equal rights under the law? Should parents who make a decision because of their religious belief be protected from suits when parents with no religious views are subjected to a suit? Medical science's ability to cope with genetic issues has increased tremendously, and new advances in science allow for fairly accurate predictions based on prenatal testing. How will these issues be decided by the courts?

Most people expect that some type of uniformity and consistency will arise from court decisions. Society can deal with consistent results, but the structure of the American court system makes uniform and consistent results unlikely. Thus, what is legal in one state may be illegal in another. I think this is precisely the kind of situation that will continue as various courts further address the subject of genetic disorders in an attempt to determine the legal issues arising from the use or misuse of genetic diagnostic procedures.

Discussion

Q It was interesting to note that in some cases where a vascetomy proved to be negligently performed and parents sued the physician on the basis that the birth of an unwanted child was due to negligence, the courts held that although such might be true, they allowed only $1 in damages. The court voted that the cost and responsibility of raising a child are offset by other, positive factors. In all cases you discussed, no reference was made to the fact that each child is a human being and is able, in many instances, to love, even a Down syndrome child, for example. What is likely to happen when the many benefits and qualities of love that children can bring to their parents and receive from them are considered when that child sues the parents for not having an abortion?

A   I think that such considerations may be introduced generally around the country when dealing with cases in which the child is normal. Generally speaking, courts permit a case to go to trial if, for some reason or another, a couple is given erroneous information on birth control pills or vasectomies and a child is conceived. They may sue the pill manufacturer or the physician or both. If they can convince the jury that they are suffering damages as a result of this error, they may be awarded compensation. The amount of compensation can vary greatly. But there is always the risk that some juries may conclude that putting the child through college, for example, may cost a great deal. Such imponderables as love and affections may be appropriate at times, and I suppose many trial lawyers would not hesitate to introduce such evidence even in a Down syndrome case. But what about a child with a fatal disease? I do not think you will find a trial lawyer talking about love and affection there. Given that type of evidence, I do not think the average trial lawyer will attempt to offset such injury with the values the child may have brought to the family in the brief periof of time he or she was healthy. This is my own evaluation of how I might be prepared to try that type of case.

Q   In the *Curlender* case, was it established that only one parent was an Ashkenazic Jew?

A   Although it was generally conceded that Mrs. Curlender has a 100 percent Ashkenazic background, it was assumed that Mr. Curlender was a Sephardic Jew. One of his grandfathers, however, had been born in Austria. It is still not known whether the grandfather was Ashkenazic or Sephardic, but this raises additional questions and leads to the realization that genetic testing, and particularly genetic screening, may be more difficult than many people realize.

Q   We decided that the determining factor in whether to set up a genetic clinic should not be the legal implications. If genetic services are part of good health care provision, then clinics should be established despite any associated legal risks. Not having the clinics does not guarantee exclusion from liability. The second point our group agreed on was that the clinics and the responsibility connected to such clinics are related to the particular level of care that is being administered by the particular hospital. In other words, if the hospital is providing tertiary care, then certainly such a clinic would be appropriate. If the hospital is providing primary-level or secondary-level care, then we would not see any great necessity to provide genetic services.

A Regional clinics have been proposed. That would be one way of making it more certain that the clinic would be staffed by really competent people.

Q Should there be mandatory screening for selected genetic diseases?

A The experts who testified in the *Curlender* case were almost universally opposed to mandatory screening. The reason they gave for being opposed was that they seriously doubted it would ultimately solve the problem. Mandatory screening would probably involve some type of governmental agency. The experts held that governmental agencies would not be able to attract the expertise that is required to make genetic evaluations and do these tests. They felt very strongly that mandatory screening would present even more problems.

Q If the parents of a young woman aged 16 or 17 years with Down syndrome, and not of normal intelligence, would like tubal ligation performed on her, could they give consent for that tubal ligation? Or, if the daughter had tubal ligation performed with parental permission, could she sue the parents later because of the tubal ligation?

A So far, only in California can the child bring such a lawsuit. Most of the urban states, however, have almost inevitably assumed that the practice of genetic medicine is really no different than any other practice of medicine. The courts simply assume genetic medicine is another area of expertise and that those who practice it are held to certain standards of expertise. It is no excuse to say "I did it wrong."

Q There are still unanswered questions. Mr. Fitzgerald, if the Down syndrome woman gives birth to a child, who is legally responsible for that child? Who can legally give permission for a Down syndrome woman to have a tubal ligation?

A As far as who is legally responsible, the parents are immediately responsible. In many states, of course, there are institutionalized possibilities, depending upon the degree of deficiency. On the second question, in California I believe that the woman could obtain permission for a tubal ligation by consultating with her own physician. If she is mentally retarded, she would have a guardian appointed by the court, and the guardian of the court would make that decision. I suspect most California judges would permit the tubal ligation.

Footnotes

1. *Curlender v. Bio-Science Laboratories,* 106 Cal. App. 3d 811, 165 Cal. Rptr. 477.

2. *Roe v. Wade,* 410 U.S. 113 (1973).
3. *In Re Quinlan,* 348 A. 2d 801, 335 A.2d 647.
4. *Turpin v. Sortini,* 119 Cal. App. 3d 690.
5. *Gleitman v. Cosgrove,* 49 N.J. 22, 227 A.2d 689 (1967).
6. *Berman v. Allan,* 80 N.J. 421, 404 A.2d 8 (1979).
7. *Park v. Chessin,* 400 N.Y.S. 2d 110.
8. *Becker v. Schwartz,* 46 N.Y. 2d 401, 386 N.E.2d 807 (1978).
9. *Howard v. Lecher,* 53 App. Div. 420, 386 N.Y.S. 2d 460 (1976).
10. *Speck v. Finegold,* 408 A.2d 496 (Pa. 1979).
11. *Gildiner v. Thomas Jefferson University Hospital,* 451 F.Supp, 692 (E.D. Pa. 1978).

• chapter 10 •

Genetic Disease, Counselors, And The Wider Society: A Philosopher's View

Gary M. Atkinson, PhD

I believe that the points I wish to make in this chapter may be best understood if placed within an autobiographical context, so I should like to begin by providing a brief account of the way in which my views have developed. I became interested in bioethics initially through a concern about abortion. I perceived that issues in the field of biomedicine were so interconnected that to deal carefully with the abortion issue, it was necessary to raise questions about the practice of medicine in general and the fundamental presuppositions underlying it. As an academician with rather scant clinical background, I worked three summers as a research associate at the Pope John XXIII Medical-Moral Research and Education Center, St. Louis. During that time I participated along with Fr. Baumiller in the Task Force on Genetic Diagnosis and Counseling sponsored by the center, and I was coeditor with Fr. Albert Moraczewski of the task force report *Genetic Counseling, The Church, and The Law.* With this theoretical background I was given the opportunity to gain clinical experience through a fellowship in clinical medical ethics at the University of Tennessee Center for the Health Sciences, Memphis. I was able to observe at three locations: counseling interviews conducted by the department of genetics of the University of Tennessee School of Medicine, a neonatal intensive care unit (NICU) associated with

---

Dr. Atkinson, is in the Department of Philosophy, College of St. Thomas, St. Paul, MN. Remarks for this paper are based partially on research that was sponsored by the Program on Human Values and Ethics at the University of Tennessee Center for the Health Sciences, Memphis, and funded by a grant from the National Endowment for the Humanities.

the medical school, and an interdisciplinary unit dealing with long-term, noninstitutionalized care of children with severe developmental problems, usually of a genetic nature.

My views on the subject of genetic counseling had been formed as part of my work on the task force report, and they were not significantly altered as a result of my clinical experience. I will offer some comments about that later, but first I wish to discuss the interrelation between genetic disease and the kind of society in which a genetic disease expresses itself.

Genetic Disease and Society

To someone unfamiliar with the field of genetic counseling, it may seem that counseling focuses primarily on the patient or client who either possesses a genetic defect or who may be at risk for having children with a genetic defect. It does not take long, however, to recognize that such an assumption is completely unjustified. Whether one observes in a clinical setting or simply examines case studies, one discovers rather quickly that the focus of counseling is not simply on the person who may be affected or possess carrier status, but on the person as he or she exists in a family. Factors that the counselor must take into account go far beyond strictly medical and genetic issues and include finances and insurance programs, the genetic history of the broader family and the familial attitudes that arise from it, the availability of particular social services, the stability of the husband-wife relationship, the existence and condition of other offspring, the possession of a job, the self-image of various family members, and, perhaps most important, the values and moral character of the persons involved. It may be something of an exaggeration, but certainly not terribly wide of the mark, to suggest that there is simply no disease, no matter how severe, that cannot be borne with grace, courage, and dignity by people whose character and familial relationships are sound and that there is no disease, no matter how mild it may appear to some, that cannot effectively destroy the persons involved. In other words, it is not simply genes that can be healthy or defective. What happens to persons as the result of genetic abnormalities is just as much a function of their values, character, and relationships with others.

That much, I think, I gleaned from my reading. What I did not realize was that the nature of the difficulty was even broader and more radical than this and that in dealing with genetic problems, we are dealing, whether we perceive this clearly or not, with issues that go beyond the patient and immediate family to the nature of society and the kind of a people we are.

The interdisciplinary unit I observed that dealt with the evaluation and care of children affected with chronic problems was known as the Child

Development Center. It was able to call upon expertise in a wide range of fields: genetics, pediatrics, neurology, dietetics, physical therapy, dentistry, orthopedics, psychology, and social work. Children were referred to the center by independent agencies. The center would perform a complete diagnostic workup of the child and conduct extensive interviews with the parents. Once a diagnosis had been made, a strategy for care would be developed and the child's progress would be followed, sometimes for years. I was able to gain information about the course of care from three sources: discussion with staff, study of interdisciplinary case reports provided by the center, and the viewing of counseling and diagnostic sessions behind a one-way mirror.

During my time at the center I saw firsthand or through case studies and discussions human suffering that made me shudder. I experienced pity and fear, but this was no drama, and there was no catharsis. What I saw were families destroyed, friends stripped away, and social life reduced to nothing. What I all too commonly saw were isolated women, women abandoned by husbands and deserted by family and friends, receiving no outside support, women who were in such a state that they were unable to leave the house to shop for groceries because they could not take their affected child with them and could find no one willing or able to babysit. I read case records stretching over years of families who would send their child to a unit of the center for what was called "respite care": the child was taken care of for a weekend or a week or two in order to permit the woman or family to relax, recover, and prepare for the next round. As I read case reports from the earliest introductory letters and general information forms to the then current evaluations and recommendations, I noted a disturbing pattern: although an initial letter would be addressed to "Mr. and Mrs. X," almost invariably it was the mother who filled out the information forms. It was she who signed them, often without the signature of the husband, even when he was presumably available to sign the forms, and I began to take this as a bad omen.

As a result of these observations, I came gradually to view genetic disease in a new light. The affected individual is the focus of attention, but I began to see the individual not as the problem, or at least not as the sole locus of the problem. Why does one person despair when others surmount the very same condition? Why do some families disintegrate while others only seem to grow stronger? And I wondered, when I saw women abandoned to their own resources, where were the grandparents, brothers and sisters, where were the friends, where were the local church and other philanthropic organizations, where were governmental agencies, where were we? I came to see that an individualistic society, a society stripped of intermediate orga-

133

nizations and groups and genuine communities, is a pretty unhealthy place for people involved with genetic disease. Environments can be unhealthy at different levels, and I would suggest that a society that does not foster, indeed, gives no thought to, the development of moral character, family life, and genuine community, is as much a cause of problems as the genetic abnormality itself.

If these points are valid, then the task confronting genetic counseling takes on an even more unhappy character. I do not mean to suggest here that it is medicine's job to right society's wrongs. But we must appreciate the radical sense in which genetic counseling can only be palliative at best. It is not simply that many genetic diseases have no medical cure. It is rather that the complex difficulty confronting those who would provide care is only partially medical in nature, and at times the medical component is not even the most troublesome one.

I think it is also important to appreciate how moral failure generates pressure for further failure. A weak marriage is somebody's fault. The absence of sustaining familial and communitarian bonds is somebody's responsibility. The isolation of a family in its need is wrong. It is not simply unfortunate, it is blameworthy; it is this isolation, the recognition that there is no one around to help, that generates pressure for "solutions" like selective abortion. For the would-be parent who knows that the full burden of care will fall on him or her alone, abortion may seem the only way out.

I do not wish to suggest here, however, that if society were more supportive of the afflicted individual, there would be no request for selective abortion, but I do believe that the perceived need would be substantially reduced. It would not be completely eliminated, though, for reasons that deserve consideration at some length.

Clinical Experience and Moral Judgment

I shall begin by raising an issue that may appear somewhat far afield: the role of clinical experience in the making of moral decisions in medicine. As an academician trained in analytic ethics, I am acutely conscious of the charge made by some physicians and others that clinical decisions in medicine cannot intelligently be evaluated morally by those who have not experienced the complexities encountered in the actual practice of medicine. I doubt that many would subscribe to a view thus baldly stated, but the suspicion regarding academicians and nonclinicians is nonetheless real and pervasive and not entirely unjustified, I hasten to add, because academicians are as subject as anyone to the temptation to develop firm convictions on topics about which they know nothing. Certainly humility and self-questioning are marks of any reasonable person, who must wonder what he or she would think if possessed of more firsthand experience. Thoughts like the following were accustomed to run through my mind: I hold certain religious views regarding the sanctity of human life, its inviolability, its inherent worth in the sight of God, regardless of its condition or status. Philosophically I believe I can show the inherently unjust, idiosyncratic, necessarily question-begging character of any judgment that would hold the lives of certain members of the human species to be of less than equal or of negative worth. But I also recognize that the inferences drawn from this position can appear quite harsh, and I wonder whether it is with some temerity that I hold views independently of firsthand experience, views contrary to those of so many whose professional work is in genetics and counseling. Where so many experienced, conscientious persons believe selective abortion and eugenic sterilization to be permissible, who am I to say them nay?

Such were my thoughts as I reflected on the interrelation between clinical experience and moral judgment. The opportunity I had to observe in a clinical setting and to talk with geneticists and counselors provided me with a different understanding of the matter, not with regard to my own moral position, but rather of the role of experience. I recognize that what I am going to relate may appear self-serving and that my conclusions have an extremely narrow basis. But I offer these reflections as a way of illustrating further complexities in the nature of genetic counseling.

In my discussions with the gentic counselors, it was several times pointed out to me that I should by all means observe some of the more severe consequences of genetic disease and that I should take a tour of a state mental institution in Arlington, Tennessee, a small town about 15 miles northeast of Memphis. I was told that there was a long waiting list to enter Arlington, and that the average IQ of its inmates was 19. "Wait till you see Arlington," I was told. Many of its inmates had been seen by the Child Development Center and had been born at the neonatal ICU in Memphis. That unit was run by an activist, pro-life director who eschewed quality-of-life judgments in decisions about whether to treat.

I did see Arlington, and I was surprised by what I saw. Or rather, I am not sure what I saw. I must confess I saw very little suffering at all; I saw little evidence of physical suffering and pain. If there was mental anguish, I did not observe signs of it. Indeed, the residents' low IQ and lack of awareness would seem to tell against strong elements of mental suffering.

Perhaps *suffering* is the wrong word to use. Perhaps *indignity* would be better. But I did not see much of that either. On the contrary, I saw a great deal of dignity. I saw many of the patients being escorted and cared for by elderly volunteers, probably bused in for the day from the large, black Baptist churches of Memphis. Now there is suffering, terrible suffering, associated with genetic disease, but it is not to be found in some of the most severely affected, and it is often to be found more in the family than in the affected individual.

My point here is simply that I did not see what others had expected me to see. "What did you go out in the wilderness to see?" asked Jesus of the crowds regarding John the Baptist. What did I go out in the wilderness of Arlington to see? And to what extent did my expectations, values, and commitments determine what I saw? Far from its being the case that experience grounds values, I am now inclined to believe that it is more often the case that values control experience. I do not mean to imply that when it comes to values everything is up for grabs and that no values can be said to be more intelligent or better grounded than others. I do mean to suggest, however, that the appeal to experience is not a simple matter and that it may be because of values that we gain the experience we do. In other words, I am less impressed by the appeal to experience than I once was, because I have come to realize how experience is so greatly filtered through expectations and commitments.

What does this have to do with genetic counseling? If my point here is correct, then I think we need to recognize the existence of fundamental, irresolvable disagreement about appropriate responses to genetic disease. For example, the person who accompanied me on my tour of Arlington said, "We cannot do enough for these children." In one sense I could agree with her remark; we cannot do all we would like to be able to do for them. But that was not what she meant. She was saying that we cannot do enough good for them to justify our keeping them alive. She was saying, in effect, that there is a certain minimum level that must be reached in order to justify the preservation of a life, and if that minimum level cannot be reached, then we should not attempt to preserve that life at all and perhaps should even take steps to end it. But how does one show, in opposition to such a view, that we *can* do enough or that there is *no* level so low that it excludes the value of any reasonable effort? I do not see how one can show this. One can show, as a matter of simple justice, that it is unfair to act on an arbitrary, idiosyncratic, and unsupported notion of what makes life worth living, at least when such action involves the killing of an unconsenting human. But even this argument presupposes that considerations of justice apply and are of overriding importance. The appeal to justice becomes particularly unpersuasive to some when attention is focused on the "suffering" or "indignity" such a position implies and when it is thought that the argument from justice is itself an imposition of arbitrary values on a couple. The existence of such irreconcilable disagreement creates obvious difficulties for a facility that is institutionally committed to a pro-life stance but must deal with sincere and committed professionals who hold conflicting values.

One other point about Arlington is worth mentioning here. I asked one of the staff what she would like to see accomplished, assuming the sky was the limit. Remembering what I had heard about the long waiting list for Arlington, I expected her to say that she would like to see another five or ten Arlingtons build in order to accommodate the many patients seeking admittance. I was quite surprised, then, when she replied that she would like to see Arlington emptied. Her view of the institution was radically different from what mine had been up until then. I had seen it as an end, a warehouse where hopeless cases were simply dumped until their death. She saw it as a waystation, a place where severely handicapped individuals could be trained sufficiently to be able to leave Arlington and be placed in homelike shelters. The problem was that an inmate would be trained sufficiently to be able to leave, but there would be no opening available for him or her. The bottleneck, as she saw it, was not Arlington but the wider community. Once again, I was struck by the social component of genetic disease in addition to its physical component.

The Role of Catholic Health Care Facilities

What I wish to do now is to take these rather general remarks and see if they have some relevance to Catholic health care facilities that offer genetic counseling and diagnostic services. In the first place, I think we can see why the practice of genetic medicine is an inherently unhappy enterprise. I hasten to admit that I have here focused on its dark side and that there are many cases where the prospective parents can be told that they are not at risk for bearing a child with a particular defect or where the child is diagnosed in utero as not affected by the disease for which it was thought to be at risk. Moreover, many types of disease are treatable, and one cannot overemphasize the hope that diagnosis and counseling will in the long run lead to fuller understanding and new cures. Nonetheless, when all the positive aspects have been assessed, and these may very well outweigh its negative side, genetic counseling still remains an unhappy enterprise because it faces difficulties it simply is not equipped to deal with. Of course, there is nothing particularly unique about this: all fields of medicine extend beyond the clinical setting and the purely physiological to the broader social context, but I think this is nowhere more painfully true than in genetics and gentic counseling. If Catholic facilities offer genetic services, they should be prepared for heartache. This is, of course, no reason for not offering them: "Surely he has borne our griefs and carried our sorrows; yet we esteemed him stricken, smitten by God, and afflicted" (Isa. 53:4). Isaiah's prophecy of the suffering servant has always by Christians been applied to our Lord, and if we are called to imitate him, we also possess the assurance of St. Paul, who could write in praise: "Blessed be the God and Father of our Lord Jesus Christ, the Father of Mercies and the God of all comfort, who comforts us in all our affliction, so that we may be able to comfort those who are in any affliction, with the comfort with which we ourselves are comforted by God. For as we share abundantly in Christ's sufferings, so through Christ we share abundantly in comfort too" (2 Cor. 1:3-5).

In the second place, if Catholic facilities do choose to offer genetic services, they should not be surprised if counselors do not share fundamental Christian values on the inherent sanctity of human life. I do not mean to justify any attitude of suspicion, and I certainly do not believe that every genetic counselor harbors un-Christian values. I will say, however, that the geneticists I had the opportunity to observe in a clinical setting seemed to me to display an uncommonly strong distaste for physical and mental abnormality. I have already suggested that it would be difficult to distinguish the extent to which this distaste was caused by exposure to the consequences of genetic disease from the degree to which some antecedent distaste was a factor in their choice of the profession of genetics. It seems to me that it is one thing to be sensitive to the pain and suffering caused by genetic disease, and it is quite another to have a horror of the condition itself. Of course, sensitivity can lead to horror and antecedent distaste can produce unusually high sensitivity.

Recommendations

I am acutely conscious that this chapter is devoid of helpful suggestions. I have been more concerned to indicate what I think to be some of the often unnoticed difficulties encountered in genetic counseling. I do so with the belief that although a better understanding of the nature of a problem is a far cry from a solution, it is nonetheless the indispensable first step. But I do have two specific recommendations to offer. First, I would strongly urge complete openness between health care facilities and counselors. Personal beliefs and institutional commitments do matter, and although counselors are sincere in their contention that counseling can be "nondirective" or "value free," it nonetheless seems to me indispensable for a sound relationship that each party be clear about the other's fundamental values. Second, I would also urge that in the presentation of the Catholic facility's position, the Church's values be expressed without apology or defensiveness. The other extreme, stridency, is also often a mark of self-doubt. Honest, forthright, respectful expression of views is possible only for those who are confident of their position, and the Church has nothing to be defensive about

Genetic Disease, Counselors, and the Wider Society: A Philosopher's View

here. Her views about genetic disease are not shared by everyone in the profession of genetics, but that is not because she is blind or insensitive. It is charity, after all, that "bears all things, believes all things, hopes all things, endures all things." It is the proclamation of this message that the Church owes to the world. And if this proclamation is foregone, her inner life is lost and she is but "sounding brass and tinkling cymbal."

Discussion

NOTE: The "Rs" represent responses from the several small groups into which the audience had been divided for discussion.

Q   How can a hospital help a couple deal with the forthcoming birth of a genetically handicapped child?

R-1   We feel that hospitals should help a family that has decided to proceed with the birth of a handicapped child through its various resources. For instance, after the child is born, the social service department could help place the child in an institution or in a home satisfactory to the family or coordinate assistance through churches and other appropriate agencies if the child is going to remain with the parents.

The department could investigate various public programs to see what monies were available and identify various nongovernmental agencies that might be of help. In general, the department could serve as a resource center for the parents and guide them to the various programs and agencies capable of providing support and help.

R-2   We feel that support should be available as soon as the parents know that their child will be genetically handicapped. Parents as well as other family members (including siblings) should receive information on what is about to happen and the changes that will take place in their lives. It is important for the siblings to understand the situation so that they do not feel slighted.

In addition to the social service department, which would provide practical assistance, the pastoral services department could provide spiritual support. In light of the current attitude of negativity toward genetically handicapped children, it is important to take a positive attitude toward the baby, emphasizing to the parents that this baby really is a gift from God, even if at first that might be very hard for them to accept.

Not to be overlooked is the concept of respite care, which gives the primary caregiver temporary relief. The hospital itself might provide appropriate referrals, or it can direct the family to an agency that offers that service. If family members know ahead of time about these services and can prepare in a reasonable manner for the birth of the handicapped child, they will be better able to cope with the situation.

Suitable training and education of in-house personnel and staff is vital, because everyone who comes in contact with these families should be able to provide help and support from the first genetic counseling session to the infant's birth.

R-3 We cannot overemphasize the importance of the pastoral staff being brought in to witness to the philosophy and values of the Catholic Church in regard to suffering. For a Christian, the value of suffering is derived from the Cross. Jesus taught us by his words and his actions that suffering is *not* punishment sent by God. Rather, it can be an occasion for a deepening of one's love of God—paradoxical as that might seem to a nonbeliever. A Catholic hospital has a wonderful opportunity to testify by its official policy and by the attitudes and actions of its personnel—at all levels—to the redemptive value of suffering. Although it may not lessen physical pain, such an attitude, when thoroughly embraced, can also help to lessen mental anguish and provide a deep faith and peace.

R-4 There is a need for community education on a broad scale to inform people that there are alternatives to abortion, that genetically handicapped children can and should be integrated into society, that offering financial and respite care to parents would be an appropriate involvement of Church groups. Families expecting a handicapped child can be placed in touch with families who have children with Down syndrome or other handicaps and have had a positive experience with that child.

Q The issue of cost-benefit weighing does not merely concern trying to save money in the health care of children or adults who need treatment; it means facing the fact that health care resources are limited. Hospitals must place resources where they will do the most good. Hence, in a cost-benefit analysis, the savings that are supposedly accrued through, for example, a screening program are shifted to an area where they can be spent to help other individuals. At stake is a weighing of one life for another life. Perhaps funds will be taken from an anencephalic child, who will probably die in the first year of life, and used to help a handicapped child who may live for many years. How does a hospital treat these situations and make such choices ethically?

A *Atkinson*: There is, I think, a cost that does not get factored into the cost-benefit analysis, and that is the cost of the attitude that holds that problems can be solved by killing human beings or, if not solved, at least ameliorated. There is no way of factoring in *that* cost. How many lives, with improved quality of life, are equivalent to the cost of that kind of attitude? There is no way of making that kind of judgment, and I suspect what happens is that that cost, because it cannot be incorporated into calculations, is simply ignored. Cost-benefit analyses are, at bottom, numerical. How can those factors be assigned in other than numerical terms or given any weight other than in a purely subjective and perhaps idiosyncratic bias? I do not see how one could possibly begin to weigh this kind of cost on a single scale. Different kinds of costs must be translated into a single dimension in order to be compared.

Q There is another closely related issue. For the past few minutes, we have been talking about a forthcoming genetically handicapped child. The parents must decide whether to have or to abort the child. For those who follow Church teachings, abortion is not acceptable. But how much effort should be made to keep a handicapped child alive, when the option of abortion does not exist? If the child is alive and has a very severe condition, such as Tay-Sachs disease in its late stage, what is the obligation of the parents, hospital, or community to keep that child alive? Does the obligation cease at any point?

A *Moraczewski*: Certain principles can be used as a guide, but first I will ask Dr. Murray to comment on anencephaly. Are there various degrees of anencephaly?

A *Murray*: Anencephaly is a birth defect in which there is grossly incomplete development of the skull associated with incomplete development of the brain with degeneration, especially of the cerebral hemispheres. These areas of the brain are seriously deficient. The face is also abnormal in structure and the ears are malformed. It appears to be an all or none phenomenon. The size of the skull defect and the degree of brain degeneration may determine how long the infant lives. They may be stillborn or die within hours or days after birth. In contrast, in persons with spina bifida there is a wide range of expression, from merely a patch of hair over the site of the defect to large segments of exposed spinal cord. In the latter instance the impairment can be extremely severe. The British have classified spina bifida patients into grades 1, 2, 3, and 4. They made a high-level policy decision not to treat children born with grade 3 or 4 spina bifida. The decision was based on cost-benefit analysis through the national health care system. The number of pounds required to treat such a child and keep that child alive is so great that the British decided that the treatment did not justify the cost. For this reason, the responsible authority instituted the alpha-fetoprotein screening program extensively.

This decision was based on the concept of triage; i.e., with limited resources—time, effort, energy, and money—treating the people who can be helped the most, since the rest would only receive minimum benefit, even with maximum effort. In other words, effort is directed to produce maximum benefit for the community. That is the basis for the above decision, as cold and harsh as it might seem. A problem arises here because a relatively small spinal defect can produce an elevation in the alpha-fetoprotein level in the amniotic fluid, which might lead to an abortion of a child whose defect is relatively minor and could be surgically repaired. This admittedly is the utilitarian approach to the problem, and, of course, from a moral standpoint, it raises significant issues.

A *Moraczewski*: In brief response to the question raised about the level of obligation to provide treatment, the principles are not complex, even if applying them to a particular situation may be difficult. First, any action should support life. Second, the obligation to maintain life is not absolute but relative. One is not obliged to use means that would be useless or would place an excessive burden on the individual. One must determine (1) whether the treatment proposed is truly a treatment, namely, that it does offer a reasonable hope of benefit and is not useless; and (2) whether the *treatment itself* would be so burdensome to the individual or others that the attainment of more important values would be jeopardized. If the treatment is useless or too burdensome, it would be considered *ethically extraordinary* or disportionate.

In the type of situation under consideration, a child born with very severe abnormalities that were *not* correctable should be given basic nursing care—food, water, and warmth. Any *treatment* that would not notably prolong the child's life or truly correct the defects or relieve possible pain would be considered as nonobligatory. Thus, a child with anencephaly who by medical judgment is deemed to be dying should be given adequate nursing care, but no heroic medical means should be instituted if they would only serve to prolong dying.

Q I would like to ask Dr. Atkinson one simple question. At the end of his lecture, he mentioned that if there were to be genetic services in Catholic hospitals, counselors should have Christian values. I think this is a reasonable suggestion; however, he also asked that counselors should be nondirective. With all due respect, how can one be a nondirective counselor and at the same time hold a specific bias? I think that there is no such thing as nondirective counseling.

A   *Atkinson*: First of all, I did not intend to say that in a Catholic hospital the counselors *should* have Christian values. Ideally, they would. I was suggesting that one might have a hard time finding such counselors because of a bias (as I perceive it) in the profession. In any event, I happen to think that nondirective counseling is possible. At first, I was unsure about it, but I have seen it work—at least, the counseling seemed to me to be nondirective. If one looks very carefully at a counselor in action, one can detect his or her attitudes, but I am not sure that the average client is that interested or that perceptive. I doubt, however, that the impact of such counseling on the individual is directive to any significant degree, since the basic values of the couple (or individual) are the result of years of living and are deeply entrenched. The likelihood that subtle messages—verbal or nonverbal—are going to alter those values or push a person to act in conflict with those basic values is extremely small.

You have asked a good question, though, about the problem of directive and nondirective counseling. I would, at least to offer a direction for a solution, suggest that there seems to be an important distinction between two kinds of counseling. One kind is counseling that involves the prospect of amniocentesis, and the other is all other counseling, the kind that Professor Baumiller showed when there is a pedigree revealing a history of genetic problems in the family. Counseling sessions involving this kind of situation seem to me to have relatively few significant ethical problems. The counselor is giving information because people are worried.

The other kind of session involves the possibility of amniocentesis, and in the sessions I have observed about 95 percent were held because of advanced maternal age with Down syndrome in mind. I believe that there is a remarkable different ethical ambiance in the two types of counseling sessions. But to answer your question directly, I believe nondirective counseling is not only possible, but desirable in the context of nonamniocentesis counseling sessions. On the other hand, when amniocentesis is involved and abortion is possible, I do see significant moral problems, which probably can only be treated adequately in the specific situation.

• chapter 11 •

Social Implications Of Genetic Manipulation

Rev. Daniel J. Sullivan, SJ, PhD

The theme of [Aldous Huxley's] *Brave New World* is not the advancement of science as such; it is the advancement of science as it affects human individuals. The triumphs of physics, chemistry and engineering are tacitly taken for granted. The only scientific advances to be specifically described are those involving the application to human beings of the results of future research in biology, physiology and psychology. It is only by means of the sciences of life that the quality of life can be radically changed.[1]

The scientific breakthroughs that Huxley spoke of in 1946 are occurring now, in the 1980s. We are witnesses to remarkable advances in the life sciences, especially in the field of genetics. In this chapter, I would like—as a priest and a biologist—to provide a balancing framework to the themes addressed in preceding chapters. In short, scientists should be free to pursue the truth through scholarly research, but society (which includes scientists) should have its rights protected.

Biological Engineering

Genetic manipulation has many meanings, depending on who is using the terms. It has also been called genetic engineering, biomanipulation, and biological engineering. Engineering can be defined as the application of sci-

---

Fr. Sullivan is associate professor in the Department of Biological Sciences, Fordham University, Bronx, NY

entific principles to practical ends. With this in mind, I am using biological engineering in a broad sense to include three levels: organismal, cellular, and molecular.[2]

Organismal Engineering

Organismal engineering involves humankind's manipulation of complete organisms whether they be plants, animals, or humans. Previous chapters have examined this topic as it relates to *Homo sapiens.* Many of the ethical and theological problems present today concern the application of humankind's ability to manipulate organisms. For instance, artificial insemination, fertilization, and embryo transfer, first used with animals, have logically been applied to humans. The technical method is essentially the same, but there is an added dimension for ethicists when artificial insemination is accomplished with sperm from the husband (AIH) or from a donor (AID), be he known or anonymous. Similarly, the problem of surrogate mothers, sperm banks, or "test-tube" babies must be faced. The case of Louise Brown, born in 1978 by in vitro fertilization, is well known; this technique has been used successfully in the United States. Even the artificial womb or laboratory chamber being studied for mammals could also be applied to humans. Such manipulation, at least as it applies to humans, also includes the entire range of surgical operations, heart transplants, skin grafts, artificial kidneys, sex change surgery, and various organ implants.

Cellular Engineering

Cellular engineering is the biomanipulation of the cells of organisms, e.g., the nucleus, chromosomes, and various cell organelles. Hence, I am excluding for the time being the genes within the chromosomes, which I will discuss under molecular engineering. Down syndrome, for instance, is a disease in humans in which there is an extra 21st chromosome. The goal of cellular engineering would be to cure the child by eliminating that extra chromosome caused by nondisjunction and resulting in trisomy. Also on the cellular level are such chromosome-related diseases as color blindness, hemophilia, and muscular dystrophy. Cloning is an example of cellular engineering and is often made the subject of science-fiction scenarios. It results in no genetic change either vertically (between parent and offspring) or horizontally (between siblings). Although practiced extensively in botany for many centuries through grafts, cuttings, and other asexual means, it is more difficult to do with higher organisms. Recently, however, some human cells called *hybridomas,* hybridized from the laboratory fusion of lymphocyte and myeloma cells, could produce clones of beneficial antibodies against cancer. There are even industrial applications: an oil-gobbling hybrid bacterium has been given a U.S. patent as a new, living, human-made microorganism.

Molecular Engineering

Molecular engineering is the rearrangement or substitution of genes on the molecular level within the chromosomes. This is true genetic engineering, although frequently the term *genetic* is used in a wider sense to include cellular and even organismal manipulation.

Although Watson and Crick published their famous discovery of the double-helix model of the deoxyribonucleic acid (DNA) molecule in 1953, 20 years passed before the next giant step. It was not until 1973 that the new technique called *gene splicing* was invented. The DNA molecule can be cut into sections, and fragments from other DNA molecules (either from the same or any other kind of organism) can then be inserted between the separated parts and reunited again into a long hybrid chain. Gene splicing thus results in *recombinant DNA*. (See Chapter 5 for further discussion.) This is molecular engineering—the fundamental basis for genetic manipulation.

Medical Applications

Recent advances in the field of recombinant DNA have been so rapid that commercial manufacturing of at least three important products—human interferon, insulin, and growth hormones—is already possible. Scientists have induced laboratory bacteria to produce these vital products of living cells. What was once in short supply and therefore extremely expensive can now be produced in greater quantity and therefore at lower cost. (See Chapters 5 and 7.)

Sometimes, however, the news media, in reporting the latest research in molecular engineering, do a disservice to the public by implying in sensational and sometimes irresponsible headlines that the ultimate panacea has been discovered. On the other hand, we can be optimistic that someday genetic manipulation will be perfected for therapeutic application. A balanced view is necessary, however. For instance, one report concerned the surprising results that use of the antiviral protein, interferon, actually had the opposite effect of what was intended. Mice infected with weak strains of the virus of lymphocytic choriomeningitis were then inoculated with interferon. Instead of reducing the pathogenic virus, as was expected, the interferon actually aggravated the virus from a docile to an aggressive form. Such research is full of surprises, and time as well as patience is required to solve these problems.

Neoeugenics

I am confident that medical applications of molecular engineering will have mainly beneficial effects for humankind. There is, however, a possible misuse of such genetic manipulation in the dangerous form of neoeugenics. This possibility was discussed by Jon Beckwith, MD, of Harvard University Medical School, when he addressed the New York Academy of Sciences.

> There has been increasing concern over the last several years that recent findings in molecular genetics and other areas of genetics are being and will be used as a means of social control.... My own fears are not based on any inherent mistrust of progress in genetics per se. Rather, they derive from an analysis of the social and political context which generates particular scientific developments and which determines their applications. This analysis reveals a society in which (1) a small number of people control the resources and power and determine the directions science takes; and (2) there is an increasing attempt to "blame the victims" of an unjust social system for society's ills. Furthermore, (3) one form this blame takes is the "medicalizing" or "biologizing" of social problems; and (4) a general attitude has been inculcated that more technology will solve our problems. Seen in this context, it is possible to foresee the application of genetics in a new eugenics movement which, in conjunction with other technologies, will be used against the poorer classes in this society in order to maintain the present power relationships.[3]

Remember that it was the Englishman, Francis Galton (1822-1911), who was the central figure in the popularization of patterns of inheritance at the family level. In 1883, he coined the word *eugenics* to describe the systematic study and application of the ideas of heredity to humans with, according to Marc Lappé, PhD, "the objective of improving (or at least stopping the degradation of) human genetic stock. His ideal of race improvement was predicated on developing a sound theoretical foundation for eugenics. But the scientific and statistical tools available at the time were unequal to the task."[4]

Eugenics must therefore be placed in its historical context—for 1883 was, of course, *after* the 1859 publication of Darwin's *The Origin of Species by Means of Natural Selection.* Indeed, natural selection was the key to Darwin's theory, and it was also the cornerstone of Galton's eugenics. Genetics was unknown at the time, however, because although the monk Gregor

Mendel orginally published his research on the genetics of peas in 1865, it was not really discovered and understood until 1900. Hence, Galton knew nothing about Mendelian genetics. Dr. Lappé, of the Hastings Center of Society and Ethics, emphasizes the point that "for the first two or three decades after the rediscovery of Mendel's laws, many scientists were optimistic that much of human misery and disease, as well as the more positive characteristics of natural ability and intelligence that Francis Galton had studied at the end of the nineteenth century, would lend themselves to simple inheritance patterns." This was genetically unscientific, and by the end of the Nazi era and World War II, eugenics fell into disrepute.

But is there not a dangerous parallel in the 1980s? Have we too been deluded into another so-called romantic period of genetic naiveté? Beckwith and others think so. There is the fear that this new tool of molecular engineering could lead to a resurgence of neoeugenics. For instance, the bioethicist, Joseph Fletcher, PhD, urges such an approach as an imperative that actively demands humankind's intervention whenever possible. In fact, he believes that not to do so would be ethically wrong. Hence, Fletcher supports situation ethics and radical opportunism when he writes:

> Our basic ethical choice as we consider man's new control over himself, over his body and his mind as well as over his society and environment, is still what it was when primitive men holed up in caves and made fires. Chance versus control. Should we leave the fruits of human reproduction to take shape at random, keeping our children dependent on accidents of romance and genetic endowment, of sexual lottery or what one physician calls "the meiotic roulette of his parents' chromosomes?" Or should we be responsible about it, that is, exercise our rational and human choice, no longer submissively trusting to the blind worship of raw nature?[5]

For Fletcher, of course, there is a moral imperative to answer in the affirmative. Other ethicists and even evolutionists, however, have serious reservations about such genetic manipulation. One problem is, Who determines the quality control, or whose ideal is being propagated? Second, from a scientific viewpoint, most evolutionists emphasize the need for variation in the gene pool rather than an arbitrary suppression of chromosomal recombination.

Hence, there is legitimate fear that molecular engineering and other new technologies could be used for social coercion in a revival of neoeugenics toward a brave new world. This seems especially conceivable today in the current value climate of secular humanism. Yet as awesome as some of these possibilities may be, there is no need to look on biological advances as necessarily opposed to humankind's well-being. Again, a balance is needed between improvement and abuse. It is important to demystify science and to recognize the fallibility of scientists. They too are human and part of society, so there should not be any dichotomy between what is good for science and what is good for society. Decisions should therefore reflect the will and needs of all people, not just the so-called experts—be they ethicists or geneticists. Society's future depends in great measure on its wisdom in using these remarkable discoveries of science. Scientists and the public must work together for the common good and avoid the dangers inherent in neoeugenics.

Asilomar and Society

Despite the good that might result from gene splicing in medicine, agriculture, industry, and even human behavior, many scientists as well as the general public have had reservations about the unintended side-effects or biohazards of this new technique. This concern was based on the use of a tumor-causing simian virus, known as SV40. SV40 was spliced into the DNA molecule of a laboratory strain of *Escherichia coli*. Unfortunately, *E. coli* is the ordinary bacterium found in the human intestine. Hence, many researchers feared that the hybridized bacterium with the SV40 tumor virus might accidentally escape from the laboratory and then invade the normal *E. coli* bacterium in humans, causing widespread disease.

Research Moratorium

In this use of recombinant DNA, the research itself (and not its application) was considered dangerous. Because of the potential biohazards involved in doing gene-splicing research, the scientists themselves imposed a moratorium on this kind of laboratory experimentation until they could meet to discuss the problem. Hence, the landmark Asilomar Conference was held in Monterey, CA, in 1975. It attracted 150 internationally renowned scientists. A remarkable concern for society and public responsibility emerged, and a majority of those attending voted to allow the National Institutes of Health to publish guidelines for future research on gene splicing. How remarkable that scientists themselves should regulate their own research! Some found it unbelievable that the sacred cow of science should be restricted, not by the Church but by the scientists themselves. The initiative of the Asilomar Conference resulted in the 1976 publication *N.I.H. Guidelines for Research on Recombinant DNA Molecules,* which had a built-in advisory committee to recommend subsequent relaxations as seemed fitting over the years.[6]

I give the historical details in order to emphasize that the scientists themselves had enough social awareness to initiate the Asilomar Conference as an almost unique event in the history of science. Daniel Callahan, PhD, supports this type of self-regulation: "Individual scientists and scientific groups are subject to the same norms of ethical responsibility as those of all other individuals and groups in society. They have neither more responsibility for their actions nor less; there is no special ethic of responsibility applying to scientists that does not apply to others."[7]

Legal Aspects

Legal aspects of genetic disorders and diagnosis, and genetic counseling and screening have been discussed in previous chapters. My purpose in mentioning the legal aspects in relation to research on molecular engineering is to dissuade the courts and politicians from getting involved. To begin with, it is doubtful that the courts will be able to contribute very much to the resolution of biohazardous research. Second, laws are made by the legislatures. Legislatures are politicans who, for personal reasons, occasionally have more than the commonweal in their purview. Third, the scientific aspects of genetic manipulation are so complicated that not even scientists agree on their effects. It would be asking the impossible of the politician, let alone the citizens represented, to make reasonable decisions on this subject. Hence, I emphasize the responsibility of self-regulation by the scientists themselves. As at Asilomar, and as is done by other professions, the scientific community must control biohazardous research within safe limits. If members fail to do so, then the public, through elected representatives and eventually the courts, will of necessity impose external restrictions. The scientists can and should avoid this. In this way, perhaps, society will have more confidence in scientists. Public fears of unintended biohazards from nuclear reactors and biomedical genetics have eroded this confidence and jeopardized traditional support of technology. Improved communication and education both between colleagues within the scientific community and above all between science and society would overcome the recent climate of distrust.

Commercialization and Universities

Advances in the field of recombinant DNA have been so rapid that commercial production of human interferon, insulin, and growth hormones is now possible. Corporations can manufacture these three results of molecular engineering, and, in fact, industrial gene splicing is now big business. It is interesting that the first ones to exploit this new technology were a few scientists and entrepreneurs rather than the major industrial giants. Spearheading this gene-splicing industry were several small companies with molecular biologists as founders and advisers: Genentech (represented today by Michael Shepard, PhD), Cetus, Genex, and Biogen. Next in the field were several large pharmaceutical companies: Eli Lilly and Company, The Upjohn Company, Pfizer Laboratories, G. D. Searle, Merck Sharp & Dohme, Hoffman-LaRoche, and Miles Laboratories. These big companies were slower to move into recombinant DNA and were also more secretive about their plans. Finally, the last but not the least arrivals on the commercial DNA scene were the major oil and chemical companies such as Standard Oil, Dupont, General Electric, and Monsanto. All have sensed the financial potential of gene splicing as a multimillion-dollar industry.

What are the social implications of this commercialization? Besides making human interferon, insulin, and growth hormones more available and at a lower cost to the consumer, there are also academic repercussions on university campuses. Industrial companies are competing so fiercely to capture what is called the "biotechnology market" that they have already signed a large number of the country's top university-based biogeneticists to lucrative contracts as consultants or board members. Some geneticists have even formed their own companies and become, at least on paper, overnight millionaires. Hence, the academic world has gradually become involved in this lucrative enterprise, and there is the danger of the "commercialization of universities." Following are several recent examples:

1. Harvard University considered a partnership in a genetic engineering company. One of its faculty members, Mark S. Ptashne, PhD, a professor of biochemistry, had done pioneering work in this field, and the proposed company would have made use of his research in conjunction with Harvard. As campus debate began, however, the faculty sharply opposed the plan. There was concern that academic freedom and the exchange of ideas and research among scientific scholars would dry up as professors came to regard their research as a trade secret. Also disturbing was the danger that granting of tenure and promotion in rank, as well as time off for company-hired professors, might be seen by colleagues as commercial favoritism. Mainly because of strong faculty opposition, Harvard's president, Derek C. Bok, PhD, rejected this proposal, citing potential conflicts of interest and possible restrictions on academic freedom. Despite the uproar at Harvard, the proposed arrangement has had close precedents at Stanford University and the University of California. After consultation with the major biogenetic engineering firms, a special "patent licensing system" was drawn up. It avoids Harvard's pitfall of going directly into partnership with biotechnological companies, but it does permit a commercial liaison.

2. A similar crisis but a much different plan has caused a problem at the Massachussets Institute of Technology (MIT). The millionaire industrialist, Edwin C. Whitehead, had dreamed for a long time of founding an institute for biogenetic research in affiliation with a university. Hence, he proposed an extraordinary arrangement between MIT and the nonprofit Whitehead Institute for Biomedical Research, to be located next to the MIT campus. Whitehead would give $20 million for the facilities, provide $5 million a year in operating funds, and leave an endowment of $100 million when he died. The institute would be headed by David Baltimore, PhD, professor of biology at MIT, and one of the pioneers in the gene-splicing technique. Opponents argued that MIT would surrender an unprecedented degree of control over the selection of research areas and over the appointment of its own faculty members. This threat to academic freedom involved the most jealously guarded rights of the faculty. The most controversial aspect of the plan was that although the institute would be administratively separate from MIT, most of its researchers would be full-fledged faculty members of the university. But despite this opposition, the plan was eventually approved.
3. Another controversy was a similar conflict of interest at Massachusetts General Hospital. Plans are being worked out to give Hoechst, the German chemical company, unduly favorable access to its biomedical research, which could result in secrecy and publication restrictions.
4. Still another controversy has arisen in which a new cell line was developed at the University of California Los Angeles (UCLA) and specimens were routinely shared with a scientist at the National Cancer Institute, Washington, DC. According to David Golde, PhD, of UCLA, a cell line capable of producing interferon was sent without his permission from the National Cancer Institute to a scientist at the Roche Institute of Molecular Biology (a nonprofit center supported by Hoffmann-La Roche, the pharmaceutical company). The Roche Institute says that in collaboration with Genentech, the biogenetics company, it used genetic engineering techniques on the cell line to develop a means of making large quantities of interferon. Envisioning commercial production, Roche applied for a patent. UCLA threatened to sue the drug company for unauthorized exploitation of its research. The drug company filed first, however, claiming that the university was interfering with its own research. The university went ahead with its suit, and now the two are locked in a titanic legal battle over the ownership of a cell line originally developed in part from a grant of federal tax dollars.
5. In another case, an out-of-court settlement was recently reached between a microbiologist, Leonard Hayflick, PhD, and the National Insti-

tutes of Health concerning proprietorship of a cell line. Many scientists were angry at the Hayflick affair in that it seemed that the government had marched into Hayflick's private laboratory and publicized what it pleased with no peer review or outside verification. An extraordinary letter appeared in *Science* supporting Hayflick and bearing the signatures of 85 prominent scientists.

These examples demonstrate the concern that many scientists and non-scientists have about the direction of genetic engineering and the basic questions about academic freedom and conflicts of interest. Sheldon Krimsky, PhD, professor of philosophy and science at Tufts University, expresses his fears as follows:

> Society needs to have neutral scientists who can address problems of whether the research is proceeding in the best interests of humanity. It would be unfortunate if no one is left to take the long view. Society is losing out if the scientists placing priorities on the research have commercial ties that lead them to keep corporate interests uppermost.[8]

Krimsky gives an excellent example. One of the new products offered as a potential benefit of gene splicing is human growth hormone, a substance that could be given to children suffering from hyposomatotropin or pituitary deficiencies, resulting in such reduced development as abnormally small size. But, Krimsky points out, the commercial market for *human* growth hormone is comparatively small compared with the much larger and more lucrative market for *bovine,* or beef-growth, hormone that would stimulate the growth of cattle. Since the research and development costs for the two products are about the same, a decision based solely on corporate boardroom business interests might well result in bringing to market a growth hormone for the beef industry before one for humans. It is mere speculation on my part, however, but may I suggest another side to the coin? Perhaps profits from the bovine growth hormone could be used in part to subsidize research and production of the human growth hormone. Am I too much of an optimist to hope for such a compromise?

Another Asilomar

Because most of the experts in the field of biomedical genetics are at the major universities, the danger of commercialization mentioned above has caused widespread concern. In a recent editorial, *The New York Times* expressed the fears of many:

> Healthy separation between pure research and its application does not at present exist in biology. The values of the marketplace have

so invaded the campus that on several occasions researchers have refused to share with their colleagues the exact details of how they did their experiments. Such attitudes are incompatible with the ethos of a scholarly community.[9]

One of the outstanding critics of the commercialization of genetic engineering at universities has been Donald Kennedy, PhD, the president of Stanford University and former commissioner of the Food and Drug Administration (FDA). He stated:

> We might begin by asking why the example of the commercialization of biomedical research is worth examining with such care. My answer is that it is probably a harbinger of things to come. Stanford and Harvard have approached this problem first, largely because no other U.S. universities have such concentrations of talent in the biomedical sciences. . . . These issues will appear not only in the area of biomedicine, but in other parts of science as well.[10]

The 1975 Asilomar Conference considered the biohazards that might result from research with recombinant DNA. Although these fears have since been reduced, the historic meeting demonstrated scientists' concern for the social implications of their research. Dr. Kennedy has now called for another Asilomar-type conference of concerned scientists to discuss the problem of commercialization at universities. He hosted such a meeting in March 1982 attended by the presidents of five major universities (Harvard, MIT, University of California, California Institute of Technology, and Stanford), leaders of ten high technology companies, and a group of prominent scientists. These 35 conference participants met near Watsonville, CA, and agreed that the quality of teaching and research should not be compromised by the growing commercialization of scientific research. Such candid discussions of mutual problems can prove to the public that scientists in universities and industry do have a social awareness based on ethical values and not merely on financial considerations.

Agrigenetics

Although this book is mainly concerned with health care, I wish to digress briefly on the topic of agriculture. As an entomologist, I am interested in the insect pests that eat our crops and especially in the use of beneficial insects as a biological control method to avoid reliance on chemical insecticides. At any rate, gene splicing can even be applied to agriculture. For instance, molecular engineering can increase the efficiency of such a basic plant function as photosynthesis, thus providing more food and oxygen for the earth.

Even grain plants might be made more resistant to cold or drought, thus increasing their geographic range for use by more people.

One of the most fascinating possibilities concerns the self-production of fertilizers. In recent years, plant breeders have gone well beyond the age-old technique of selecting or hybridizing new and special types of fruits and vegetables. Thanks to the "Green Revolution" of Nobel prizewinner, Norman Borlaug, PhD, superwheats and superrices have been produced.

Unfortunately, these supergrains do not naturally have nitrogen-fixing bacteria within their roots, as is the case with the well-known leguminous plants, e.g., alfalfa, clover, and soybeans. Instead, these supergrains require extra amounts of chemical fertilizers, which must be artificially synthesized. This is expensive, and the chemical industry's charge for this service is so costly as to make the Green Revolution prohibitive for the very developing nations that most need it. Through genetic engineering, however, it may be possible to transfer the nitrogen-fixing, or "nif" gene from leguminous plants to grain plants like wheat, corn, and rice. Hence, food would be provided for a hungry world as the plants themselves produced their own fertilizer.

The Pursuit of Balance

As I mentioned earlier, it is fundamental to my thinking that although disagreements exist on certain topics among scientists, and between scientists and society, agreement on most other issues is possible. I am convinced that intelligent, educated people of good will are capable of reconciling their differences and striking a *balance* in their common pursuit of truth. Society must respect scientists' freedom to do scholarly research, and scientists must realize their ethical responsibilities to society. Neither group can live any longer in isolation. This is a very exciting period of human history, and I am optimistic that the social implications of genetic manipulation will be understood and properly implemented.

Discussion

NOTE: The Q & A involved a panel of previous speakers.

Q   I address the following question to all the panelists: Why is there a fear of technological advances in general and of genetics in particular? Is it the fear that everyone has about something new? Is it the fear of living in a society (Western culture) that is losing some of its drive? Barbara Jackson, the economists, and others have postulated that because U.S. citizens live in a pluralistic society, they have no common bottom line on which to base themselves. Is the fear essentially internal, i.e., inside society? Or is the fear outside of it, concerning things that society or the individual cannot handle? That is the question I would like to ask, because I believe fear colors many of the questions surrounding the field of genetics.

A *Sullivan*: I think the fear comes from the public, which frequently does not know the issues or misunderstands them. I have no fear at all of scientists, who are dedicated, honest people trying to do their best in seeking the truth about the world. I think the Asilomar Conference on recombinant DNA technology demonstrated that scientists on their own can do something to regulate their research.

A *Moraczewski*: At first, as has been intimated, there may be a fear of the unknown because there is much that society and scientists do not yet know about the ramifications of genetic engineering. The second source of fear is the fear of losing control over the technology or its product. In this connection, one thinks of the "monster" created by Frankenstein that ran amok. In recent experience there is nuclear energy and the Three Mile Island incident. Part of the fear of nuclear power has been magnified by the media; they have focused on the sensational, which fed a common fear based on the misconception that a nuclear plant would blow up as if it were an atomic bomb. Apparently, then, the fears associated with nuclear energy and genetic engineering are fears of loss of control: human beings have created something they *cannot* adequately control.

Comment: I think another fear prevails in society today—the fear of the expert. Perhaps society fears the reputed narrow vision of the expert who knows well the details in one field but fails to perceive the "big picture" and consequently can make serious errors in judgment.

A *Packman*: One can posit still another fear. I am reminded that at one time research implicating the XYY karyotype as being associated with particular patterns of behavior was subject to opposition and attempts at suppression not only because the research itself might have been noxious to the subjects involved, but also, I believe, because people were afraid of what the findings might be: that particularly human behavior might be influenced by the same biological and chemical processes and constraints that govern biological phenomena in general. I think philosophy and religion should allow people to accept the consequences of such research, namely, that biochemical processes are involved in some human attributes, rather than to fear these findings.

A *Moraczewski*: For a while it was suggested that the XYY karyotype was associated with an increased tendency to criminality. The observation was made that imprisoned men had a higher percentage of the karyotype than did men in the general population. Hence, it was concluded that, somehow, criminal tendencies for increased aggressiveness were associated with that particular chromosomal pattern. But further and more careful research has shown that the karyotype is not necessarily associated with criminality; rather, it is associated with increased height.

Q What is the mechanism by which society monitors research without destroying its free expression, and, in particular, how would it monitor genetic engineering?

A *Shepard*: The rein society as a whole has over science is a very basic one because citizens, through Congress, control all federal money. The government has regulations for the involvement of human subjects in research. It organizes panels that either accept or reject applications for research monies. In addition, in the private sector, corporations that fund research, both in universities and in other corporations (e.g., the relationship between Hoffmann-LaRoche and Genentech) are controlled by consumers through the marketplace. If there is no public interest in this sort of research, then it will not be done. On the other hand, an imaginative, creative scientist guards with his life the freedom to pursue an idea that he views as interesting and vital. The concerns raised by scientists working in the area of recombinant DNA were not the first concerns expressed by scientists with doubts about their own research. Charles Darwin, for instance, delayed publishing his observations for a very long time because he was aware of the impact that his research would have on the whole world.

A *Sullivan*: One result of the Asilomar Conference was that an overwhelming majority of participants suggested to the National Institutes of Health that guidelines be formulated and published. The conference was held in 1975, and in 1976 the NIH *Guidelines for Research Involving Recombinant DNA Molecules* was published by the then United States Department of Health, Education, and Welfare. These guidelines set up different biological and physical safety regulations regarding this research. Also set up was the Recombinant Advisory Committee (RAC). Over the years, the RAC has modified these guidelines because the biohazards feared in 1975 never really materialized. A serious limitation of these guidelines was that the only groups forced to abide by them were those receiving federal tax monies, namely, universities and others. Therefore, pharmaceutical companies—and they are the major research groups in the field—were outside the guidelines. Although these companies voluntarily elected to follow the guidelines, there was no ready way to enforce them. To date, genetic horror has not yet been loosed out of the laboratory. Because the research community took the guidelines seriously, they have been relaxed considerably in the past year, when the hazards proved to be much less than anticipated.

Q Do you recommend a different form of monitoring than that presently used? Do you recommend that something be done?

A *Sullivan*: No, I trust the scientists, and if something blows up, such as in the Three Mile Island incident, then somebody investigates it. It is more dangerous, I believe, to allow politicians to draw up unwise regulations for scientific research, since that can stifle important work.

A *Packman*: The question has three components. 1. Should there be regulation of the *intellectual process?* I believe there should be no such regulation other than the peer review process that already exists in science, law, medicine, and every other professional field. The quality of work is reviewed by people who can best make the necessary judgments. The intellectual activity itself should not be regulated. 2. Should the research *activity* be regulated some way because it is dangerous? This question has, in my view, been settled. The research activity in itself is not dangerous, and there need be no regulation apart from self-regulation. Scientists regulate themselves every day. They regulate the use of radioisotopes in laboratories because, frankly, they fear cancer and other radiation-induced disease. They regulate the use of infectious agents because they do not want to contract serious infections. They monitor the performance of experiments on human subjects because they too are human beings. 3. Should the *application* of research be regulated? The answer must be "yes." The mechanisms for that are already present in the form of patent laws, tax laws, various business laws, congressional power to regulate interstate commerce, and congressional, legislative, and judicial power to rule in favor of public safety. I believe these mechanisms have worked reasonably well in all fields of technology to date, and I do not see any fundamental difference in their application to recombinant DNA technology.

A *Shepard*: I would like to comment on industry not being obliged to follow the guidelines. Genentech was clearly one of the leaders in this field, and it voluntarily complied with all the NIH guidelines. Genentech sent to Donald S. Frederickson, MD (then director of NIH), full descriptions and intended application of all experiments. He responded by approving all of them. It is that kind of precedent that provides a basis for the trust of continued self-regulation. A kind of peer pressure is applied on the other industries to also comply.

A *Moraczewski*: What appears to be a new phenomenon in the industrialization of scientific research is that fact that many scientists may reap large financial benefits by being founders and owners of the new companies. Relative to the regulation of science, Pope John Paul II made an interesting comment when addressing the Pontifical Academy of Science on the occasion of Albert Einstein's centenary. He distinguished between pure science and applied science. Pure science includes astrophysics, mathematics, and many parts of chemistry and biology. In these instances, the scientist pursues knowledge to find out what *is,* and the process does not involve human subjects. Scientific research in these instances can be considered unrestricted. The only measure, the only control, is truth—reality itself. Applied science, on the other hand, involves the *application* of knowledge, and when the application involves human beings or when human subjects are part of experiments, then it is essential to consider the rights of the human subjects and the impact of the research on human beings. To that extent, that research process would be limited, not primarily by an external imposition of regulations, but by the intrinsic conditions of the experiment, namely, that humans as subjects or objects, are involved in the research process.

Footnotes

1. Aldous Huxley, *Brave New World,* 2nd ed. (New York: Harper & Row, 1946). p. xii.
2. Daniel J. Sullivan, "Biological Engineering," *Thought* 56 (1981): 199-211.
3. Jon Beckwith, "Social and Political Uses of Genetics in the United States: Past and Present," in *Ethical and Scientific Issues Posed by Human Uses of Molecular Genetics,* Marc Lappé and Robert S. Morison, eds. *Annals of the New York Academy of Sciences* 265 (1976):46-58.
4. Marc Lappé, *Genetic Politics: The Limits of Biological Control* (New York: Simon and Schuster, 1979).
5. Joseph Fletcher, *The Ethics of Genetic Control* (Garden City, NY: Anchor Books, 1974).
6. Daniel J. Sullivan, "Gene-Splicing: The Eighth Day of Creation," *America* 137 (1977):440-443.
7. Daniel Callahan, "Ethical Responsibility in Science in the Face of Uncertain Consequences," in *Ethical and Scientific Issues Posed by Human Uses of Molecular Genetics,* Marc Lappé and Robert S. Morison, eds. *Annals of the New York Academy of Sciences* 265 (1976):1-12.
8. Sheldon Krimsky, "Tinkering with Life," *Science '81* 2 (1981):45-49.
9. Editorial, "Pure Science and Pure Profit," *The New York Times,* Nov. 16, 1981.
10. Donald Kennedy, "Business, Science and the Universities," *Committee for Corporate Support of Private Universities, Inc.* (1981):8-12.

Epilogue

Sr. Betty Brucker, SSM

The interest that called for this book to be published, the expertise that these chapters reflect, and the promise that they hold for the future of medicine and genetic counseling have a proper place in today's Christian-Catholic health care setting. Catholic health care professionals and facilities cannot stand on the sidelines while others develop the scientific knowledge and the clinical experience at the present explosive pace.

What is done in the way of genetic experimentation and how research is applied depends on the philosophies and value systems of the people involved and the institutions in which they work. Nothing grows in a vacuum. It is time for all involved in the healing ministry of Christ to see to it that this relatively new field of genetic medicine, with its vast potential for good, be given a chance to develop in the atmosphere the Catholic value system can provide.

If Catholic health care facilities do not provide now for the legitimate needs of research and the ever-increasing number of people who require genetic screening and other help this field of knowledge can provide, Catholics will be forced to watch people who share their values go elsewhere for help and see research and a field of medicine develop in circumstances often foreign to Catholic values. This must be avoided if Catholic health care professionals are to be faithful to their mission and of service to an ever-growing number of people who have legitimate need of genetic medicine.

No matter what is undertaken—whether it be a procedure in diagnostic radiology or nuclear medicine or genetic research and experimentation—there must be a basic emphasis on a living out of the Christian values for which Catholic health care was founded. Those values permeate whatever is done, and, above all, there will be intellectual curiosity and intellectual competence. There is always a desire to serve God's people in the most complete and in the most advanced ways possible.

I think amniocentesis and genetic counseling are the major causes of hesitancy in making genetic medicine a fully accredited service at a Christian-Catholic setting. For many, these two terms are simply fancy words for

---

Sr. Brucker is executive director, St. Mary's Health Center, St. Louis, Mo.

abortion. Without denying that the information amniocentesis provides is sometimes used to justify abortion, it is nonetheless a scientifically safe procedure, the performance of which has many sound medical reasons: Rh studies for the Rh factor and prematurity studies; detection of chromosomal defects such as mongolism or Down syndrome; and the preparation of parents for possible birth disorders, to name but a few.

It is in this last area that the Catholic value system can make its most telling contribution. By helping the couple expecting an abnormal baby cope with this experience; by showing them that there are people who care; by helping them develop their own support group, Catholic health care professionals make their pro-life stance a reasonable alternative to abortion. Unless Catholics become active in this field, they will have little to say to the growing number of couples for whom amniocentesis and genetic counseling are medically indicated, and they will have little impact on other major developments in the genetic medicine and genetic engineering "explosion."

Allow me to share the experience of Saint Mary's Health Center of St. Louis. This health center is most fortunate in having had a cytogenetics laboratory for the last 10 years. Sr. Leo Rita Volk, SSM, director of this laboratory, whose interest in genetics and belief in its importance began in the mid 1950s, made sure staff at St. Mary's were aware of developments in this field and the potential it held for the future. But even when it became obvious that genetic counseling was the logical next step in developing a genetic medicine program, it was not an easy step to take. Besides the usual administrative difficulties, major problems were finding a genetic counselor whose values reflected those of the Sisters of Saint Mary and overcoming a fear that some individuals and groups would consider the health center to be encouraging abortion.

When St. Mary's found a qualified genetic counselor, it still had to face the fear of possible misunderstanding from staff, other Christian-Catholic facilities, and community prolife groups. Theologians and genetic authorities, however, met with physicians, supervisory personnel, and pro-life groups and released well-designed statements to the press, preventing much of this. The program was put in place with little adverse criticism and very few of the anticipated problems. I share this to underline the possibility that these fears were unfounded or at least exaggerated and that people of good will respond to the truth.

People need what genetic medicine offers now and what it will offer in the future. Even more important, people and the whole field of genetic medicine need the moral context in which the Christian-Catholic community offers these services. By acting now, the Catholic health care system will be a party in shaping the future of genetic medicine. Values are never easily achieved and never easily preserved. There will be need for constant vigilance over genetic and counseling services provided; there will be need for vigilance over the sense of responsibility and accountability demanded of personnel involved. The quality of people associated with this important work will be a very important factor in ensuring that despite risks and dangers, a course respectful of human life and human values can be charted. The challenges of these dynamic and energetic undertakings require value-centered people to meet them. Catholic health care facilities have such people.

Glossary*

ACHONDROPLASIA. An autosomal dominant disease involving an abnormality in the conversion of cartilage into bone, resulting in a severely dwarfed individual.

AID. Artificial insemination with donor semen.

AIH. Artificial insemination with the husband's semen.

ALLELES. Alternative forms of a gene. If more than two alleles exist for a given locus, they are called multiple alleles—for example, all the mutant genes at the hemoglobin beta-chain locus.

AMINO ACIDS. The building blocks of proteins. Each has an amino group on one end, a carboxyl group on the other, and a side group (R) that gives it its specificity.

AMNIOCENTESIS. A procedure in which a needle is inserted through the abdomen of a pregnant woman into the amniotic sac in order to remove some of the fluid. Tests done on the amniotic fluid and on fetal skin cells suspended therein allow a number of genetic disorders to be diagnosed and the maturity and condition of the fetus to be determined.

AMNIOGRAPHY. A procedure for visualizing the outline of the placenta and umbilical cord by injecting a radiopaque dye into the amniotic sac and subsequently obtaining an x-ray film of its contents.

AMNIONITIS. An infection developing within the uterine cavity of a pregnant woman.

ANENCEPHALY. A condition in which brain tissue and skull are absent to a variable extent.

ANTHROPIC PRINCIPLE. The concept proposed by several physicists (e.g., Dicke and Carter) that the universe as it actually exists can be best understood in the light of human presence.

ANTIBODY. A gamma globulin formed by immune-competent cells in response to an antigenic stimulus and reacting specifically with that antigen.

ANTIGEN. A substance having the power to elicit antibody formation by

---

*The definitions of many of these terms are based on those found in Nora, James J. and Fraser, F. Clarke, *Medical Genetics: Principles and Practice* (Philadelphia: Lea & Febiger, 1981) and appear here with the publisher's permission.

165

immune-competent cells and to react specifically with the antibody so produced.

ARTIFICIAL INSEMINATION. The procedure by which semen is deposited by a suitable device, such as a syringe, close to the cervix or into the uterine cavity.

AUTORADIOGRAPHY. A technique whereby the precise location of a radioactively labeled molecule in a cell or tissue can be demonstrated by applying a photographic emulsion to the histological section or cytological slide; the film will be sensitized wherever the label is present. Applied in cytogenetics, particularly to delineating deoxyribonucleic acid (DNA) synthesis by the chromosome by adding tritium-labeled thymidine to the culture, the label will be incorporated wherever DNA synthesis is proceeding.

AUTOSOMAL DOMINANT INHERITANCE. A genetic condition that requires for its manifestation that only one of a paired genes (alleles) is a mutant.

AUTOSOMAL RECESSIVE INHERITANCE. Refers to a condition or trait that is manifest or expressed only when each of a paired somatic chromosomes carries the same gene.

AUTOSOME. Any chromosome other than the sex chromosomes.

BIOMANIPULATION. A general term encompassing a variety of technologies such as genetic engineering, reproductive technologies, and cloning in which life forms are produced or modified by human art.

CARRIER. An individual who carries a gene but may not manifest it, i.e., either an autosomal or X-linked recessive gene or a dominant mutant gene that has not yet resulted in overt disease.

CELL DIFFERENTIATION. The process by which cells with identical genetic composition nonetheless gradually begin to differ from one another in the same multicellular organism, apparently due to a selective repression of certain genetic traits or characteristics.

CHIMERA. An individual composed of cells derived from different zygotes; in human genetics, especially used with reference to blood group chimerism, a phenomenon in which dizygotic twins (i.e., nonidentical) exchange hematopoietic stem cells in utero and continue to form blood cells of both types. To be distinguished from mosaicism, in which the two genetically different cell lines arise after fertilization.

CHOREOATHETOSIS. Dancelike abnormal movements of the body.

CHROMOSOMES. The carriers of the genes, consisting of long strands of DNA in a protein framework. The exact structure of mammalian chromosomes is still not known. In nondividing cells they are not individually distinguishable in the nucleus, but at mitosis they become condensed into visible strands that stain deeply with basic stains.

CHROMOSOME TRANSLOCATION. At one point during cell division, the chromosomes are duplicated by a complex biochemical process. These products are shared equally by the daughter cells. At times the chromosome duplication/separation goes awry, however, perhaps due to an as yet unidentified toxin, and a portion of a chromosome gets attached to another chromosome. The error is perpetuated in the progeny.

CITRULLINEMIA. An autosomal recessive disorder of amino acid metabolism in which there is abnormally high concentration of citrulline in the blood, cerebrospinal fluid (CSF), and urine, resulting in mental retardation.

CLONE. A group of cells all derived from a single cell by repeated mitosis and all having the same genetic constitution (in somatic cell genetics).

CODON. A triplet of three nucleotide bases in a DNA or messenger ribonucleic acid (RNA) molecule that codes for a specific amino acid or the initiation or termination of transcription.

CONSEQUENTIALISM. A form of a teleological ethic that morally evaluates an action by only its consequences.

CONSULTAND (consultee). The person, in a genetic counseling situation, whose genotype is being evaluated—often the parents of an affected child.

COOLEY'S ANEMIA. Another term for thalassemia major. It is one of several hereditary anemias involving a reduced production of hemoglobin and results in an enlarged spleen, growth retardation, bone changes, and, frequently, death in childhood.

CYCLOPIA. A congenital defect in which the two eye sockets are fused, resulting in only one eye.

CYSTIC FIBROSIS. An autosomal recessive disease, characterized by abnormally thick secretions that tend to obstruct the organ ducts, resulting in a variety of clinical problems.

CYSTINOSIS. An autosomal recessive disease involving an inborn error in the metabolism of the amino acid cystine, which leads to the deposition of cystine crystals in several systems, including kidneys and eyes.

CYSTINURIA. A condition found in certain inheritable diseases (such as cystinosis) in which there is an excessive amount of cystine in the urine.

CYTOGENETIC TECHNOLOGY. The laboratory procedures by which the chromosomes of a cell are made visible and photographed. The pairs of chromosomes are then arranged according to a standard pattern to form the karyotype.

DEONTOLOGISM. The group of ethical theories that holds that moral right and wrong is determined, ultimately, by obedience to some code of laws or customs.

DEOXYRIBONUCLEIC ACID (DNA). A macrolinear molecule made up of four deoxyribonucleotides in a variable sequence that carries genetic information and is found in the chromosomes of all cells with a nucleus. The genetic material of all eukaryote cells.

DEOXYRIBONUCLEOTIDE. A compound that consists of a purine or pyrimidine base bonded to the sugar 2-deoxyribose, which in turn is bound to a phosphate group.

DIPLOID CELL. A cell having the full complement of chromosomes for that species, which for the human is 46.

DIZYGOTIC (dizygous, fraternal). Type of twins produced by two separate ova, fertilized by separate sperms.

DOMINANT. A gene is said to be dominant if the phenotype of the heterozygote is the same as that of the homozygote for that gene. In human genetics the term is used, more loosely, for a mutant gene that is expressed in the heterozygote. If the mutant homozygote is more severely affected than the heterozygote, there is "intermediate" dominance, and if both genes are expressed independently, there is "codominance." Traditionally, the term refers to traits, but is now commonly applied to genes as well.

DOWN SYNDROME. A congenital disease resulting from the individual possessing in the cells of the body (in the most common variety) three chromosome no. 21 (rather than the normal two), which is manifested by a variety of abnormalities including varying degrees of mental retardation, retarded physical growth, flat face, thickened tongue, and stubby fingers.

DRIFT, genetic. Chance variation in gene frequency from one generation to another. The smaller the population, the greater are the random variations.

DYSPLASIA. Abnormal development of tissue.

E. COLI. A species of the bacterium *Escherichia*. It occurs normally in the human intestine as well as in other vertebrates.

EMBRYO TRANSFER. The process by which a very young embryo, usually resulting from in vitro fertilization, is placed in the uterine cavity of the same or other female from whom the egg has been obtained.

EMOTIVISM. A form of noncognitive deontologism that holds that a person has an obligation to follow his or her own natural and psychologically healthy feelings in deciding what is a morally good or evil action.

ENCEPHALOCELE. A protrusion of brain tissue through an opening in the skull.

ENDONUCLEASE. An enzyme that makes internal cuts at specific sites in DNA backbone chains.

ENZYMES. Protein molecules capable of catalyzing specific chemical reactions.

EUGENICS. The science that deals with the improvement of the human race by careful selection of those who are to reproduce.

EUKARYOTIC CELL. A cell in which the nuclear material is enclosed in a membrane so as to form a distinct nucleus.

EXPRESSIVITY. The variability in the degree to which a mutant gene expresses itself in different mutant individuals.

FAMILIAL TRAIT. A trait that occurs with a higher frequency in the near relatives of individuals with the trait than in unrelated individuals from the same population.

FERTILIZATION. Fusion of gametes of opposite sex to produce a diploid zygote, i.e., a cell with the full complement of chromosomes.

FETAL ALCOHOL SYNDROME. A condition occuring in a fetus who has been exposed to relatively high and continuing levels of ethyl alcohol in the maternal blood stream. Among the physical characteristics found in infants born with this condition are short upturned nose, thin upper lip, indistinct philtrum (groove under the nose), possible mental retardation, and subtle defects in learning skills.

FETOSCOPY. A procedure involving the use of fiberoptics system inserted into the amniotic sac with a needle, permitting direct viewing of the fetus. At present, the risks to the fetus are still high, about 3 to 5 percent, even when done by experienced persons.

GALACTOSEMIA. An autosomal recessive disease in which, because of the absence of an enzyme, the sugar galactose is not properly metabolized and thus accumulates in the tissues.

GANGLIOSIDOSIS. An autosomal recessive disease characterized by the abnormal accumulation in the nervous system of certain gangliosides (combinations of lipids and various carbohydrates).

GENE. The basic unit of heredity by which specific traits or characteristics are transmitted from one generation to the next. Most genes are located in chromosomes in the nucleus (for organisms with nucleated cells). Chemically, genes are segments of a DNA molecule and code for the amino acid sequence of a particular polypeptide chain.

GENE SPLICING. The general procedure by which a DNA segment—representing a desired gene—is "spliced" or inserted into the DNA molecule of another cell, usually of a different species.

GENETIC CODE. The sequence of bases (guanine, cytosine, adenine, and thymine) in the DNA molecule that carries the (genetic) information for the synthesis of amino acids. Each amino acid is represented by a unique sequence of three bases.

GENETIC ENGINEERING. The science and art concerned with altering the specific genetic makeup of a cell or organism.

GENETIC MANIPULATION. A general term that can encompass genetic engineering (see above) and gene splicing (see above).

GENOME. The collection or set of genes found in an individual.

GENOTYPE. The genetic constitution of an individual, with respect either to (1) his or her complete complement of genes or to (2) a particular locus. Contrast with PHENOTYPE.

HAPLOID CELL. A cell having only one half the full complement of chromosomes for a member of that species. Normally, the mature reproductive cell—the egg and the sperm—are haploid.

HEREDITARY, HERITABLE, HEREDOFAMILIAL. Essentially synonymous terms for genetic traits. Formerly, *hereditary* was sometimes used in the sense of dominant. *Heredofamilial* is archaic.

HERMAPHRODITE. An individual with both ovarian and testicular tissue (not necessarily functional).

HETEROZYGOTE. An individual in whom the corresponding locus in each of paired chromosomes is occupied by different genes.

HETEROZYGOUS (HETEROZYGOTE). Possessing different alleles at a given locus. Double heterozygote refers to a heterozygous state at two separate loci. An individual heterozygous for two mutant alleles, such as those for hemoglobin S and C, may be called a compound heterozygote.

HOLOPROSENCEPHALY. An abnormal condition where the brain has failed to develop into two hemispheres.

HOMOCYSTINURIA. An autosomal recessive disease involving a defective enzyme formation and characterized by the urinary excretion of homocystine, mental retardation, dislocated eye lens, sparse blond hair, and tendency toward convulsions.

HOMOLOGOUS CHROMOSOMES. Chromosomes that pair during meiosis, have essentially the same morphology, and contain genes governing the same characteristics.

HOMOZYGOTE. An individual in whom the corresponding locus on site in each of paired chromosomes is occupied by the same gene.

HOMOZYGOUS (HOMOZYGOTE). Possessing identical alleles at a given locus. Contrast HETEROZYGOUS.

HOMOZYGOUS MUTANT GENE. A gene that has undergone an alteration (mutation) and is found in both members of a pair of chromosomes.

HOMOZYGOUS WILD-TYPE GENE. The normal or natural type (allele) of a rare mutant gene.

HYBRIDOMA. A hybrid cell obtained by fusing an antigen-sensitized lymphoid cell (from spleen) with a myeloma cell. The resulting hybridoma can be cloned to produce extremely pure monoclonal antibodies.

HYDROCEPHALY. A condition in which there is an excessive accumulation of fluid in the cerebral ventricles, resulting in a thinning out of brain tissue and a separation of cranial bones.

HYDRONEPHROSIS. A condition in which there is a dilation of the kidney pelvis (the basin-shaped cavity in which the urine drains) due to an obstruction to the free outflow of urine.

HYPOSOMATOTROPISM. A condition in which there is a deficiency in the excretion of somatotropin (growth hormone).

IGHD. Inherited Growth Hormone Deficiency.

IMMUNOGLOBULIN. Protein molecule, produced by plasma cell, that recognizes and binds a specific antigen. Also called ANTIBODY.

INBORN ERROR OF METABOLISM. A condition that results from the congenital lack of a specific functional enzyme, leading to a variety of biochemical effects, the results of which can be manifested in a host of clinical consequences. Presently more than 200 disorders are known, of which a dozen or so may be treated with fair or good success.

INCOMPLETE DOMINANCE. A term used sometimes as a synonym for intermediate dominance (see DOMINANT) and sometimes to refer to a mutant gene that is expressed in some heterozygotes and all homozygotes.

INFRAHUMAN LIFE. Any corporeal life form other than human.

INTERFERON. A small protein produced by cells in response to infection by virus particles that is capable of inducing resistance to additional infection by related or unrelated virus.

171

INTUITIONISM. An approach to ethics that holds that moral principles can be formulated by nondiscursive methods that permit one to grasp them directly by a direct awareness of the values involved.
IN VITRO (Latin: in glass). Refers to experiments done on biological systems outside the intact organism. Contrast IN VIVO.
IN VITRO FERTILIZATION. The process by which the sperm (human or otherwise) fertilizes a human (or other) ovum outside of a living body in some suitable container.
IN VIVO. (Latin: in life). Refers to experiments done in a system such that the organism remains intact. Contrast IN VITRO.

KARYOTYPE. The chromosome set of an individual. The term also refers to photomicrographs of a set of chromosomes arranged in a standard classification.

LEROY DISEASE (syndrome). An autosomal recessive disease characterized by growth and mental retardation, stiff joints, congenital dislocation of the hip, kyphosis, and other skeletal abnormalities.
LESCH-NYHAN SYNDROME. A sex-linked disease involving an abnormality in uric acid metabolism and central nervous system (CNS) function, resulting in hyperirritability, spasticity, choreoathetoid movements, and mutilation of lips and fingers after teeth have erupted.
LOAD, GENETIC. The sum total of death and disease caused by mutant genes.
LYMPHOCYTIC CHORIOMENINGITIS. A virus infection marked by lymphocytic infiltration of the choroid flexuses and a cerebral meningitis.

MANIC-DEPRESSIVE PSYCHOSIS. A major affective disorder in which the person can experience marked mood changes from hyperelation and activity to profound depression. The condition is marked by a tendency to remissions and recurrences with varying time intervals.
MAPLE SYRUP URINE DISEASE. An autosomal recessive disease so named because the urine smells like maple syrup. The disorder is apparently related to a deficiency in a particular enzyme that results in the elevation in blood and urine of the levels of three amino acids: valine, leucine, and isoleucine. Untreated patients (children) die within a year following progressive neurological deterioration.
MECKEL-GRUBER SYNDROME. An autosomal recessive disease characterized by kidney dysplasia, polydactyl, and occipital meningoencephalocele.

MEIOSIS. The special type of cell division by which gametes, containing the haploid number of chromosomes, are produced from diploid cells. Two meiotic divisions occur. Reduction in number takes place during meiosis I.

MENINGOENCEPHALOCELE. A protrusion of membranes and brain tissue through an opening in the skull.

MENINGOMYELOCELE. A protrusion of the membranes and spinal cord through an abnormal opening in the vertebral column.

MESSENGER RNA. The ribonucleic acid that carries the specific genetic information from the DNA in the nucleus to the cytoplasm, where it serves as a template for protein synthesis.

METHYLMALONIC ACIDEMIA. The presence of large amounts of methylmalonic acid in the blood. A recessive trait disease manifesting mental retardation, growth deficiency, recurrent episodes of nausea and vomiting, acidosis, coma, and death in early infancy.

MICROCEPHALY. A condition wherein the head is abnormally small, with a skull capacity of less than 1350 cc.

MITOSIS. Somatic cell division resulting in the formation of two cells, each with the same chromosome complement as the parent cell.

MONOZYGOTIC (monozygous, identical). Refers to twins derived from one egg and thus genetically identical.

MUCOLIPIDOSES. Autosomal recessive inherited conditions involving abnormal storage in the viscera of mucopolysaccharides and glycolipids, resulting in varying degrees of mental retardation and skeletal abnormalities.

MULTIFACTORIAL. Determined by multiple genetic and nongenetic factors, each making a relatively small contribution to the phenotype. See also POLYGENIC.

MULTIFACTORIAL DISEASES. Disorders that are the result of the interaction of several genes with environmental factors.

MUTAGENS. Chemical substances that are capable of altering the structure and function of a gene.

MUTANT GENE. A gene that has undergone a change in its chemical structure so as to alter its activity.

MUTATION. A permanent change in genetic material. The term usually refers to point mutation, i.e., change in a single gene, but in a more general sense includes the occurrence of chromosomal aberrations. In connection with inherited diseases, mutation in the germ cells is most relevant, but somatic mutation also occurs and may be important in relation to neoplasia and aging.

MYOTONIC DYSTROPHY. Autosomal dominant disease, slowly progressive usually after onset in the third decade and marked by atrophy of the muscles, failing vision, slurred speech, and muscular weakness.

NANISM. Another term for dwarfism—from the Latin, *nanus* (dwarf).

NATURAL MORAL LAWS. Those norms regarding human behavior that are based on a reasoned analysis of nature and the basic needs common to all members of the human species.

NEURAL TUBE DEFECT. A condition, probably multifactorial in causation, in which during embryogenesis the neural tube fails to close properly, resulting in such disorders as spina bifida and anencephaly.

NEUROFIBROMATOSIS. An autosomal dominant disease whose diagnostic features include small, discrete pigmented skin lesions (cafe-au-lait spots), multiple subcutaneous tumors often following the course of a nerve trunk, and occasionally tumors of the brain or spinal cord. There is marked clinical variability in the symptoms.

NUCLEOTIDES. The basic structural units of DNA and RNA molecules. They are composed of a molecule of phosphoric acid, a molecule of the sugar deoxyribose (for DNA) or ribose (for RNA), and one of four organic bases: adenine, guanine, cytosine, and thymine (for DNA) or uracil (for RNA).

NUCLEIC ACID. A nucleotide polymer. See also DNA and RNA.

NUCLEOSIDE. The combination of a purine or pyrimidine base and a sugar.

PATHOGEN. Any substance—chemical, viral, microbial, or other—that causes a disease.

PENETRANCE. The percentage frequency with which a heterozygous dominant, or homozygous recessive, mutant gene produces the mutant phenotype. Failure to do so is called *nonpenetrance,* and penetrance less than 100 percent is *reduced penetrance.*

PHENOTYPE. 1. The observable characteristics of an individual as determined by his genotype and the environment in which he develops. 2. In a more limited sense the outward expression of some particular gene or genes. Thus a heterozygote and homozygote for a fully dominant gene will have the same phenotype, but different genotypes.

PHENYLKETONURIA (PKU). An autosomal recessive disease in which, due to a deficiency of the enzyme phenylalanine hydroxylase, there is an accumulation in the tissues, including the brain, of phenylalanine, resulting in altered brain function, namely, mental retardation.

PIERRE ROBIN SYNDROME. Also termed Robin syndrome, this condition of undetermined cause is characterized in the newborn by an abnormally small jaw, falling tongue (glossoptosis), and respiratory distress.

PLASMID. The extrachromosomal hereditary unit that, as found in certain bacteria (e.g., *E. coli*), is a circular segment of the DNA molecule.

POLYDACTYLY. A condition where either the hand or foot has more than five digits.

POLYGENIC INHERITANCE. The determination of a trait or condition by multiple genes as distinguished from single gene inheritance.

POLYMER. A regular, covalently bonded arrangement of basic subunits (monomers) produced by repetitive application of one or a few chemical reactions.

POLYMERASE. A class of enzymes whose general function is to catalyze the formation of larger molecules from smaller units. DNA P. catalyzes the assembly of the deoxynucleotides to form the DNA molecule. RNA P. catalyzes the assembly of the ribonucleotides, using a strand of DNA as a template, to form a molecule of RNA, which thus carries the genetic information from the nucleus to the cytoplasm.

POTTER SYNDROME. A rare condition in which the face is flattened, often with widely spaced eyes and low-set, floppy ears. There is renal agenesis or other severe kidney abnormality.

PROBAND (propositus). The affected family member through which the genetic pedigree of a family is ascertained—the index case. Originally a proband was not necessarily affected and a propositus was, but by current usage the terms are synonymous.

PROPORTIONALISM. A contemporary system of teleological ethics that holds that a human act is judged to be morally good or evil depending solely on the proportion of values or dysvalues in the total concrete situation of that act. In theory, at least, this approach eliminates the possibility of there being absolute moral norms, i.e., norms that admit of no exceptions.

PRUDENTIAL PERSONALISM. A method of moral decision making based on firm commitment to the self-actualization of human persons in relation to God and community as the ultimate goal of human life and on an evaluation of alternate means of achieving this goal according to an objective and reasoned analysis of human nature and human needs in light of their potential impact on self and others.

RECOMBINANT DNA TECHNOLOGY. The group of procedures by which a gene—chemically represented as a segment of DNA—is transferred to the DNA of another organism where it can carry out its directive functions.

REPLICATION. The process by which multiple copies of an entity, e.g., a virus particle or a DNA molecule, are made. Most precisely, replication should be distinguished from *biological reproduction* by which an organism *duplicates itself.* This is in contrast to replication where an entity *is duplicated* by the anabolic activites of an organism.

RESTRICTION ENZYME ANALYSIS. A biochemical method of determining chromosomal structure by studying the pattern of fragments produced when chromosomal DNA is exposed to restriction enzymes (which have the property of cleaving the DNA molecule at specific sites).

RECESSIVE. Refers to a trait that is expressed only in individuals homozygous for the gene concerned. Usage now justifies applying the term to the gene as well. The definition is an operational one—whether a "recessive" gene is expressed in the heterozygote may depend on the means used to detect it.

RECOMBINATION. The formation of new combinations of linked genes by the occurrence of a cross-over at some point between their loci.

REDUCTION DIVISION. The first meiotic division, so called because at this stage the chromosome number per cell is reduced from diploid to haploid.

RIBONUCLEOTIDE. A compound that consists of a purine or pyrimidine base bonded to ribose, which in turn is esterified with a phosphate group.

RIBOSOMES. Granules of ribonucleic acid which are involved in the cellular synthesis of proteins.

RNA (ribonucleic acid). A nucleic acid formed upon a DNA template and taking part in the synthesis of polypeptides. It is found in the nucleus and cytoplasm of all cells and variously involved in the transfer of genetic information from the gene (a segment of DNA) to the synthesis of protein molecules. Instead of thymine, RNA contains uracil. Three forms are recognized: (1) messenger RNA (mRNA), which is the template upon which polypeptides are synthesized; (2) transfer RNA (tRNA or sRNA, soluble RNA), which in cooperation with the ribosomes brings activated amino acids into position along the mRNA template; and (3) ribosomal RNA (rRNA), a component of the ribosomes, which function as nonspecific sites of polypeptide synthesis.

SCHIZOPHRENIA. A group of mental disorders characterized by varying degrees and frequency of hallucinations and delusions accompanied by a withdrawing from people and the outside world with a concurrent focusing on an inner or private world.

SEX CHROMOSOMES. Chromosomes responsible for sex determination. In human beings, the X and Y chromosomes.

SEX-LINKED. Determined by a gene located on the X or Y chromosome. Since most sex-linked traits are determined by genes on the X chromosome, the term is often assumed to refer to these; X-linked is the preferable term in such cases.

SMITH-LEMLI-OPITZ SYNDROME. An autosomal disease that clinically includes several symptoms: failure to thrive, mental retardation, microcephaly, low-set ears, small lower jaw, short neck, and webbing between the second and third toes.

SPASTIC PARAPLEGIA. Paralysis of the lower extremities associated with spasmodic contractions of the muscles.

SPINA BIFIDA. A congenital condition, multifactorial as to mode of inheritance, in which there is a limited defect of the spinal column, involving a protrusion of spinal membranes, with or without spinal cord tissue.

SV 40. Tumor-causing simian virus, formerly used in recombinant DNA research.

SYNDROME. A characteristic association of several anomalies in the same individual, implying that they are causally related.

TAY-SACHS DISEASE. An autosomal recessive disease characterized by severe developmental retardation leading to blindness, dementia, paralysis, and death, usually by age 2 or 3 years.

TERATOGEN. A substance that can bring about abnormal development, especially in the fetus.

THALASSEMIA. An anemia due to an inherited disorder of hemoglobin production in which there is a partial suppression of one of the globin chains making up the hemoglobin molecule.

THALASSEMIA MAJOR. The thalassemia resulting from a homozygous state of the relevant gene (see COOLEY'S ANEMIA).

TRANSCRIPTION. The process whereby the genetic information contained in DNA is transferred by the ordering of a complementary sequence of bases to the messenger RNA as it is being synthesized.

TRANSCULTURALISM. An alternative term for PRUDENTIAL PERSONALISM, which emphasizes that moral norms should transcend the mores and values of particular cultures.

TRANSFER RNA (tRNA, sRNA). Any of at least 20 structurally similar species of RNA, all of which have a molecular weight 25,000. Each species of RNA molecule is able to combine covalently with a specific amino acid and to hydrogen-bond with at least one mRNA nucleotide triplet.

TRANSLATION. The process whereby the genetic information present in an mRNA molecule directs the order of the specific amino acids during protein synthesis.

TRISOMY 18. An abnormal condition in which the cells of the individuals have three (rather than two) copies of chromosome no. 18. The clinical features of this condition, which seem to affect females more than males (3:1), include mental and growth retardation, small chin, low-set malformed ears, congenital heart anomalies, rocker-bottom shaped feet, and urogenital malformations.

TURNER SYNDROME. A disorder in which the individual has only 45 chromosomes, having only one of the sex chromosomes, an X. Clinical features include dwarfism, webbed neck, infantile sexual development, and amenorrhea.

ULTRASONOGRAPHY. A noninvasive technique involving the use of high frequency sound by which the fetus and placenta can be visualized in vivo.

UTILITARIANISM. An ethical system based on the principles of acting so as to achieve the greatest good for the greatest number of people; the most favored form of it today is sometimes called consequentialism, which proceeds on a kind of cost-benefit calculation, either according to general norms (rule utilitarianism) or without such norms (act utilitarianism or situationism).

VOLUNTARISM. The ethical system that holds that a behavior is right or wrong because so determined by the will of God, of a human law maker, or the good will of the autonomous individual.

WILD TYPE. The normal allele of a rare mutant gene, sometimes symbolized by +.

WRONGFUL BIRTH. A term applied to a cause of action where the *parents* allege that the unwanted birth and related responsibilities of parenthood caused them to suffer economic loss and mental and physical pain.

WRONGFUL LIFE. A term applied to a cause of action where the *child* (or the parents on his behalf) alleges that his birth resulted in his having to bear a life filled with mental and physical pain.

ZYGOTE. The diploid cell that results from fertilization, i.e., fusion of the pronuclei of sperm and egg; it is the first cell of the new individual. The fertilized ovum or, more loosely, the organism developing from it.

Bibliography

The books and articles listed below were recommended by the speakers at the conference, Genetics and Health Care. The publishers, however, do not necessarily agree with the value judgments made in these works. Nonetheless, these references are of use for anyone who desires an initial understanding of the relevant literature. Additional references may be found in *Genetic Counseling, The Church and The Law.*

Scientific References

Anderson, W. French, "Gene Therapy," *Journal of the American Medical Association,* 246 (Dec. 11, 1981): 2737-2739.

"An Assessment of the Hazards of Amniocentesis: Report of the MRC Working Party on Amniocentesis," *British Journal of Obstetrics and Gynecology,* (1978), 85, suppl. 2.

Angier, N. "The Organic Computer. Tomorrow's Microchips May be Built of Proteins and Manufactured by Bacteria." *Discover,* 3 (1982): 76-79.

Chamberlain, Jocelyn, "Human Benefits and Costs of a National Screening Program for Neural-Tube Defects," *The Lancet,* (Dec. 16, 1978): 1293-1297.

Chedd, G. "Genetic Gibberish in the Code of Life." *Science 81,* Vol 2 (1981): 50-55.

Davidson, E. H. *Gene Activity in Early Development,* 2d ed. New York: Academic Press.

Erbe, Richard W., "Principles of Medical Genetics," *The New England Journal of Medicine,* 294 (Feb. 12 & 26, 1976): 381-383; 480-482.

Golbus, Mitchell S., et al., "Prenatal Genetic Diagnosis in 3000 Amniocentesis," *New England Journal of Medicine,* 300 (Jan. 25, 1979): 157-163.

Gottlieb Duttweiler Institute for Economic and Social Studies, *Genetic Engineering,* New York: Interbook, 1975.

Hodgen, Gary D., "Antenatal Diagnosis and Treatment of Fetal Skeletal Malformations," *Journal of the American Medical Association,* 246 (Sept. 4, 1981): 1079-1083.

Judson, H. F. *The Eighth Day of Creation.* New York: Simon and Schuster, 1979.

Kaback, M. M., "Prenatal Diagnosis of Hereditary Disease and Congenital Defects," *Pediatric Annals,* 10 (February 1981).

Kelly, Thaddeus E., *Clinical Genetics and Genetic Counseling,* Chicago: Year Book Medical Publishers, 1980.

Khorana, H. G., "Total Synthesis of a Gene," *Science,* 203 (Feb. 16, 1979): 614-625.

Lehninger, A. L. *Biochemistry,* 2d ed. New York: Worth Publishers, 1975.

Lewin, R. *Gene Expression. 1. Bacterial Genomes.* New York: John Wiley and Sons, 1978.

McKinnell, Robert G., *Cloning: A Biologist Reports,* Minneapolis: The University of Minnesota, 1979.

McKusick, Victor, *Mendelian Inheritance in Man,* 5th edition, Baltimore, MD: Johns Hopkins, 1978.

Mercola, Karen E. and Martin J. Cline, "The Potentials of Inserting New Genetic Information," *New England Journal of Medicine,* 303 (Nov. 27, 1980): 1297-1300.

Milunsky, Aubrey, *Genetic Disorders and the Fetus, Diagnosis, Prevention and Treatment,* New York: Plenum Press, 1979.

National Academy of Sciences, *Genetic Screening, Procedural Guidance and Recommendations,* Washington, DC: National Academy of Sciences, 1975.

National Academy of Sciences, *Genetic Screening, Programs, Principles, and Research,* Washington, DC: National Academy of Sciences, 1975.

National Academy of Science, *Research With Recombinant DNA,* Washington, DC: National Academy of Sciences, 1977.

National Institute of Child Health and Human Development, *Antenatal Diagnosis,* Report of a Consensus Development Conference. Bethesda, MD: National Institute of Health, April 1979, p. 4-6, 33-91.

Nora, J. J. and F. Clarke Fraser, *Medical Genetics: Principles and Practice,* 2nd Ed., Philadelphia: Lea and Febiger, 1981.

Novick, R. P. "Plasmids." *Scientific American,* 243 (1980): 103-127.

Omenn, Gilbert S., "Prenatal Diagnosis of Genetic Disorders," *Science,* 200 (May 26, 1978): 952-958.

Phillips, III, J. A., *et. al.* "Molecular Basis for Familial Isolated Growth Hormone Deficiency," *Proceedings of the National Academy of Sciences,* 78 (1981): 6372-6375.

"The Risk of Amniocentesis," *The Lancet,* (December 16, 1978): 1287-1288.

Stanbury, J. B., J. B. Wyngaarden, and D. S. Fredrickson, *The Metabolic Basis of Inherited Disease,* 4th edition, New York: McGraw-Hill, 1978.

Thompson, J. S., and M. W. Thompson, *Genetics in Medicine,* 3rd edition, Philadelphia: Saunders, 1980.

Vogel, F., and A. G. Motulsky, *Human Genetics: Problems and Approaches* New York: Springer-Verlag, 1979.

Watson, J. D. *Molecular Biology of the Gene,* 2d ed. New York: W. A. Benjamin.

Wetzel, R. "Applications of Recombinant DNA Technology," *American Scientist,* 68 (1980): 664-665.

Ethical, Legal, and Social References

Adelberg, E. A. et al. "The Frankenstein Monster and Recombinant DNA." *Hospital Practice,* 14 (1979): 214

Anderson, W. French and John C. Fletcher, "Gene Therapy in Human Beings: When Is It Ethical to Begin?" *New England Journal of Medicine,* 303 (Nov. 27, 1980): 1293-1297.

Barry, OP, Robert, "Ethics and the Christian Genetic Counselor," *Priest* 37 (Feb. 1981): 47-49.

Boone, C. K. "Recombinant DNA and Nuremberg: Toward the New Application of Old Principles." *Perspectives in Biology and Medicine,* 23 (1980): 240-254.

Capron, Alexander M., "Tort Liability in Genetic Counseling," *Columbia Law Review,* 79 (May 1979): 618-648.

Capron, Alexander M., et al., eds, *Genetic Counseling: Facts, Values, and Norms,* New York: Alan R. Liss, 1979.

Curtis, III, Roy. "Genetic Manipulation of Microorganisms: Potential Benefits and Biohazards," *Annual Review of Microbiology* 30 (1976): 507-533.

Desposito, Franklin, "Prenatal Testing and Genetic Counseling: Physician's Responsibility," *Hospital Progress,* 60 (Feb. 1980): 81.

Eibach, Ulrich, "Genetic Research and A Responsible Ethic," *Theology Digest,* 29 (Summer 1981): 113-117.

Friedman, Jane M., "Legal Implications of Amniocentesis," *University of Pennsylvania Law Review,* 123 (1974): 92-156.

Gordon, Hymie, Letter to Editor, *A.L.L. About Issues,* (Feb. 1983): 4-5.

Grobstein, C. *The Double Image of the Double Helix: The Recombinant DNA Debate.* San Francisco: W. H. Freeman, 1979.

Hilton, Bruce, et al., *Ethical Issues in Human Genetics,* New York: Plenum Press, 1974.

Kelly, OP, Bishop Thomas, Claire Randall, and Rabbi Bernard Mandelbaum, "The Control of New Life Forms—An Interreligious Statement of General Secretaries," *Origins,* 10 (1980): 98-99.

Lappé, Marc and Robert S. Morison, eds., *Ethical and Scientific Issues Posed by Human Uses of Molecular Genetics,* Annals of the New York Academy of Sciences, New York: New York Academy of Sciences, Vol. 265, 1976.

Lebel, SJ, Robert Roger, "Ethical Issues Arising in the Genetic Counseling Relationship," in *Birth Defects:* Original Article Series, White Plains, NY: The National Foundation-March of Dimes, Vol. XIV, No. 9, 1978.

Lipkin, Mack and Peter T. Rowley, eds., *Genetic Responsibility,* New York: Plenum Press, 1974.

Lisson, SJ, Edwin L., "Patenting Life Forms and Preserving Human Values," *Hospital Progress,* 62 (Jan. 1981): 36-40.

McCormick, Richard A., "Genetic Medicine: Notes on the Moral Literature," in his *Notes on Moral Theology 1965 through 1980.* Washington, DC: University Press of America, 1981. pp. 401-422.

McCormick Richard A., "Genetic Engineering" in his *Notes on Moral Theology 1965 Through 1980,* Washington, DC: University Press of America, 1981, pp. 278-329.

McFadden, Charles J., *The Dignity of Life,* Huntington, IN: Our Sunday Visitor, 1976.

Malonev, R., "Some Notes on Genetic Engineering," *Linacre Quarterly,* 47 (Aug. 1980): 266-273.

Millard, Charles E., "The Effects of Modern Therapeutics on the Human Gene Pool," *Rhode Island Medical Journal,* 63 (Nov. 1980): 443-450.

Milunsky, A., and G. J. Annas, eds., *Genetics and the Law*, New York: Plenum Press, 1976.

Milunsky, A., and G. J. Annas, eds., *Genetics and the Law II*, New York: Plenum Press, 1980.

Monteleone, Patricia L., and James A. Monteleone, "Genetic Unit Offers Diagnosis, Counseling," *Hospital Progress*, 61 (October 1980): 46-51.

Monteleone, Patricia L., and Albert S. Moraczewski, OP, "Medical and Ethical Aspects of the Prenatal Diagnosis of Genetic Disease," in *New Perspectives on Human Abortion*, eds., Thomas W. Hilgers, Dennis J. Horan, and David Mall, Frederick, MD: University Publications of America, 1981.

Moraczewski, OP, Albert S., "Some Moral Dimensions in Genetic Counseling," *Hospital Progress*, 61 (Oct. 1980): 52-55.

Moser, Mary Beth, "Genetic Counseling: An Ethical and Legal Problem," *Law Reports*, 7 (August, 1982): Appendix, St. Louis: Catholic Health Association.

Nelson, Robert J., "Genetic Engineering: Federal Ethics Commission Hears Theologian," *Hospital Progress*, 63 (December 1982): 42-47.

Neumann, Marguerite, ed., *The Tricentennial People*, Ames, IA: Iowa State University Press, 1978.

Nolan-Haley, Jacqueline M., "Amniocentesis and Human Quality Control," *The Human Life Review*, (Spring, 1982): 51-67.

Pope John Paul II (Selected writings and addresses on Ethics and Science)

Redeemer of Man, March 4, 1979, English translation, Washington, DC: United States Catholic Conference.

Address to Executive Council of UNESCO, June 2, 1980, English translation, *ORIGINS* 10 (June 12, 1980): 58-64.

Address to Italian physicians and surgeons, October 27, 1980. English translation, *ORIGINS*, 10 (Nov. 13, 1980): 351-352.

Address of Pontifical Academy of Sciences, October 3, 1981. English translation, *L'Osservatore Romano*, (Oct. 12, 1981): 4ff.

Address to participants in a study week on biological experimentation sponsored by the Pontifical Academy of Sciences, Oct. 23, 1982, *ORIGINS* 12 (Nov. 4, 1982): 342-343.

Address to representatives of Spain's Academic Community, Nov. 3, 1982. English Translation, *ORIGINS,* 12 (Nov. 18, 1982): 369-371.

Powledge, Tabitha M., and John Fletcher, co-directors, "Guidelines for the Ethical, Social and Legal Issues in Prenatal Diagnosis," A Report of the Genetics Research Group of the Hastings Center, *New England Journal of Medicine,* 300 (1979): 168-172.

President's Commission for the Study of Ethical Problems in Medicine and Biomedical and Behavioral Research, *"Splicing Life,"* Washington, DC: U.S. Government Printing Office, 1982.

Ramsey, Paul, *Fabricated Man, The Ethics of Genetic Control,* New Haven, CT: Yale University Press, 1980.

Reich, Warren T., ed., *Encyclopedia of Bioethics,* New York: The Free Press, 1978.

Reilly, Philip, *Genetics, Law, and Social Policy,* Cambridge, MA: Harvard University Press, 1977.

Rensberger, Boyce, "Tinkering with Life," *Science 81,* 2 (Nov. 1981): 44-49.

Rosner, Fred, "Recombinant DNA, Cloning, Genetic Engineering, and Judaism," *New York State Journal of Medicine,* 79 (Aug. 1979): 1439-1444.

Shannon, Thomas Anthony, "Ethical Implications of Developments in Genetics," *Linacre Quarterly,* 47 (Nov. 1980): 346-368.

Shannon, Thomas Anthony, "A New Design for Life: Ethics and the Genetic Revolution," *New Catholic World,* 224 (May/June 1981): 111-113.

Siegel, Seymour, "Genetic Engineering." *Linacre Quarterly,* (February 1983): 45-55.

Singer, M. et. al. "What Lessons Does the Recombinant DNA Debate Teach Us: A Round Table Discussion." in *Recombinant DNA and Genetic Experimentation,* J. Morgan and W. J. Whelan eds. New York: Pergamon Press, 1979.

Sullivan, SJ, Daniel J., "Gene Splicing: The Eighth Day of Creation," *America,* (Dec. 17, 1977): 440-443.

Sullivan, SJ, Daniel J., "Agromedicine: Ecological Basis for Ethical Concern," *Linacre Quarterly,* 46 (Nov. 1979): 362-366.

Sullivan, SJ, Daniel J., "Recombinant DNA: The Debate Goes On," *America,* (Nov. 10, 1979): 280.

Sullivan, SJ, Daniel J., "Biological Evolution," *Thought,* 56 (June, 1981): 199-211.

Turner, J. Howard, T. Terry Hayashi, and Donald D. Pogoloff, "Legal and Social Issues in Medical Genetics," *American Journal of Obstetrics and Gynecology,* 134 (May 1, 1979): 83-99.

Wade, Nicholas, "UCLA Gene Therapy Racked by Friendly Fine," *Science,* 210 (Oct. 31, 1980): 509-511.

Walters, LeRoy, ed., *Bibliography of Bioethics,* New York: The Free Press, 1975-1982.

Weiss, Joan O. "Religion & Genetics: The Possible Effect of Good Pastoral Care on Genetic Counseling Clinics," *Social Thought,* 8 (1982): 2-7.

Yin, K. "Recombinant DNA: Biotechnology Becomes Big Business." *Science for the People,* 12 (1980): 5-7.

Index

For the benefit of the reader this Index contains the combined indexes from this book, *Genetic Medicine and Engineering*, and *Genetic Counseling, The Church and the Law*, published by The Pope John XXIII Medical-Moral Research and Education Center. The page numbers for index entries from *Genetic Counseling* are in bold type.

Abnormal chromosomes, 3, 4
Abnormal gene, 3
Abortion, 131
   alternatives to, **151**
   arguments against, **104-107**
   arguments for, **101-104**
   and consequentialism, 90-91
   emotional response to, 89
   ethical issues on, 96-98
   as issue in genetic counseling, 95
   and natural moral law, 92-93
   obligation of the State to oppose, **104**
   and proportionalism, 94
   selective, **20, 23-24, 105-107**
   slides of, 96
   Supreme Court and decisions on, 88
Acetylcholinesterase, 56
Achondroplasia, 7, 38
Agriculture, 155-156
Agrigenetics, 155-156
Alpha-fetoprotein
   and amniocentesis, 25, 43
   in genetic screening, 55-57
   and neural tube defects, **14, 16, 47**
   tests for, **14**
Alpha-thalassemia, 50

Amino acids, 65, 70, 107
Ammonia in citrullinemia, 70
Amniocentesis, 9, **15,** 19-26, 39, 43, 82
   benefits of, **22, 23, 127-129**
   to the fetus, **22, 23, 127**
   in Catholic health care facility, 96-98, **147-151**
   decision for, 40
   ethical issues and, 143
   indications for, 19-25
   justification of, **20, 23, 127-129**
   leading to abortion, **20, 21, 23**
   miscarriage associated with, 25-27
   procedure for, **15**
   risks for, **16, 18, 20,** 25-26
   safety of, **15, 18**
   and sex selection, **20, 23-24**
   technique for, 19
Amniography, 18
Amnionitis, 26
Anencephaly, 141; *see also* Neural tube defects
   diagnosis of, with ultrasonography, 17
   in spina bifida, 35
Antenatal diagnosis
   application of, to genetic screening, **20-23**

application of, to sex selection, **23-24**
benefits of, to fetus, **19**
concealment of,
   to protect marriage, **225-228**
   to protect normal child, **228-230**
current status of, **12-15**
future prospects of, **15-17**
indirect methods of, **15**
risks of, **18**
techniques of, **13-15**
and value assessment, **18-20, 148**
Anthropic principle, 106-107
Antiepileptic drug, 9
Argininosuccinic acid synthetase, 69
Artificial insemination, 146
Asilomar, 150
Autosomal dominant disease, **7, 8**
Autosomal recessive disease, 7
   autosomal dominant, **7, 8**
Autosomal recessive gene, 5-6
Autosomes, **6**
Ayer, A. J., 89

Bacteria, xi
   genetic alteration of, 107
Balanced translocation, 21-22
Berman v. Allen, 125
Biohazards, 150
Biological engineering, 145-147
Biotin, 70
Biotin-responsive multiple carboxylase deficiency, 71-73
Birth defects; *see* Congenital malformations
Bishops and genetic disease, **143-146**

Boyle, J., 87
Branched-chain amino acids, 70
Brave New World, 145

Caiaphas principle, 90
Carbohydrate, 66
Carboxylase enzymes, 71-72
Care, standard of, **201-204**
Carrier detection, tests for, 7
Catholic health care facility
   amniocentesis in, 96-98, **147-151**
   ethical problems for, 96-98
   and genetic counseling, **146-151**
   role of, 137-138
Cause of action in "wrongful birth" cases; *see also* Legal issues on genetic diagnostic procedures
   arguments against, **183-187**
   arguments for, **182-183, 195-198**
   limited, **187-195**
Cell differentiation, 4
Cell division, 4
Cellular engineering, 146-147
Cerebral palsy, 32
Choreoathetosis, 7
Christian community as church, **141-143**
Christian genetic counselor, **111-112, 132-138,** 142-143
Christian health professional, **131-132**
Christian personalism, **92-97**
Christian witness
   individual, **113-139**
   institutional, **141-151**

Chromosomal disorders, 7-8
  frequency of, 11
Chromosome abnormality, 15
  and diabetes, 28
  genetic screening for, 55
  and maternal age, 11, 19-20, 28
  nondisjunctional, 19-20
Chromosome methods, 9
Chromosome translocation, 21
Chromosomes, xi, 4, **6**
  abnormal, 3, 4
Church, the, 93
  and genetic disease, **143-146**
  and health care facilities; see
    Catholic health care facility
Citrullinemia, 66, 69-70
Civil society; see Society
Cleft lip and palate, 8
Clinical experience and moral
  judgment, 134-137
Cloning, 75, 146
Code of Hammurabi, 88
Cofactors, 70
Coleman v. Garrison, **186-187,
  188, 189**
Commercialization of
  recombinant DNA, 151-154
Confidentiality
  breach of, **222-225**
  and duty to inform, **208, 222,
    225-230**
  legal aspects of, **134-135,
    222-230**
  moral aspects of, **40, 134-135**
Congenital malformations, 4
  causes of, 5-9
  heart defects, 7, 8
    associated with teratogens, 9
  hypothyroidism, genetic
    screening for, 50-51
Consent, informed, **208-211**

Consequentialism, 89-90, 90-91
Constitution of United States, 87
Contraception
  and Church teaching, 95,
    118-119, 125
  and natural moral law, 92-93
Cooley's anemia, 53
Cost-benefit analysis
  of abortion, **103, 106**
  in genetic screening, 46
Cost-effectiveness of health care,
  140-142
Counseling; see Genetic
  counseling
Counselors, genetic; see Genetic
  counselor
Curlender case, 121-124, 127,
  128-129
Cyclopia, 17
Cystic fibrosis, **7, 34-36**
  and pedigree, 34
Cystinuria, 64-65
  therapy for, 66

Darwin, C., 148
Deafness
  hereditary, legal issues on,
    124-125
  associated with teratogens, 9
Declaration on Abortion, 94
Declaration of Euthanasia, 94
Declaration of Independence, 87
Defective genes, 102
Demurrer, 122
Deontologism, 91
Deoxyribonucleic acid (DNA), xi,
  5, 76, 107; *see also*
  Recombinant DNA
  technology
  molecule, 4
  sequencing, 80

189

Dilantin; *see* Diphenylhydantoin
Diphenylhydantoin, 9
Directive counseling, **26,** 58
Disclosure of information
　legal aspects of, **205-208**
　moral aspects of, **39-40,
　　129-131**
Diseases; *see also* Genetic
　diseases
　caused by teratogens, 8-9
Divine providence, 106, 109
DNA; *see* Deoxyribonucleic acid
Doctor-patient relationship, **27**
Doe v. Bolton, **20, 156, 193, 194,
　221**
Dominant gene, 7
Dominion, 109
　over nature, 104-107
Down syndrome, 8, 146
　and amniocentesis, 43
　genetic screening for, 55
　incidence of, **10**
　karyotype in, 21-22
　legal issues on, 124, 126, 129,
　　206
　maternal age and, **23,** 19-20,
　　49-50, 55, **124, 148**
　mental retardation in, 20
　prenatal diagnosis of, 55
　recurrence of, 22, **50**
Duchenne's disease, **8,** 24
Dwarfism, diagnosis of, with
　ultrasonography, 17

Egg cells, 28
Electrocardiography, **13**
Electrophoresis, 54
Emotivism, 89
Encephalocele, 6
　amniocentesis for, 25

Enzyme deficiency, 7, 12
Enzymes, 4
　HGPRT, 6
Eschaton, 116-117
Escherichia coli, 76-78, 150
Ethical issues
　and amniocentesis, 143
　genetic manipulation, 101-119
Ethical principles and genetic
　medicine, 87-99
Ethics, 87
　definition of, 88
　priority of, over technology,
　　116
Eugenicism, 97, 148
Evolution, 101, 106
　human control over, 107-109
Exceptionism, 90
Exterminative medicine, 96
Extra fingers, 6

False-positive results, 45
Familiaris Consortio, 95
Family
　as support system, 133-134
　according to Vatican II,
　　116-117
Federal programs for the
　handicapped, **159-161**
Fetal alcohol syndrome, 9
Fetal sonography, 38-39; *see also*
　Ultrasonography
Fetal trauma associated with
　amniocentesis, 26
Fetoscopy, **13-14,** 18
　trauma with, 28-29
Fetus, as members of human
　family, 93
Fibroblasts, 72

Index

Food and Drug Administration (FDA), 56-57
4/21 translocation, 21
Free will, 113

Galactosemia, **21,** 50, 66
Galton, F., 148
Gene splicing, 147
Gene therapy, xii
Genes, xi
  abnormal, 3
  chromosomes, **5-7**
  and disease, **7-10**
  dominant, 7
  isolation and characterization of, 75
  mutant, 4
  nitrogen-fixing, 156
Genetic code, xi
Genetic counseling, 6, 9, 31-44, 63, 131-143
  directive, **26,** 58
  ethics in, 96, 97
  focus of, 132
  and law, **28-29, 40-41**
  legal duty to provide, **204-208**
  modification of, 117,
  and moral values of counselor, **135-138**
  nature of, **25-29**
  nondirective, **26-27,** 142-143
Genetic Counseling, The Church and The Law, 95, 131
Genetic counselor; *see also* Christian genetic counselor
  function of, **25-27**
  moral responsibilities of, **27, 28, 111, 132-138**
  legal responsibilities of, **28-29, 177-231**
  values of, 142-143

Genetic defects, 96; *see also* Genetic diseases
Genetic diagnosis, 15-29, 95-96, 98
Genetic diagnostic procedures, legal issues relating to, 121-130
Genetic diseases, xii, 3-13
  and abortion, **1, 20, 21, 23, 28**
  autosomal recessive, 7
  case histories of, **29-50**
  and Catholic health care facilities, **146-151**
  chromosomal imbalances in, 7, 9
  federal legislation, **159-161**
  genetic engineering for diagnosis of, 76
  impact of, 15-16
  multifactorial traits, **7, 8**
  number of, **6, 10, 11**
  and parents, **110-111**
  and society, **112**
  and teaching Church, **143-146**
  treatment of, 63-74
  types of, **7-12, 29-30**
  untreatableness of most, **21, 22**
  X-linked, **7-8**
Genetic dwarfs, 82-84
Genetic engineering, xi, 75-84, 101, 145
  challenge of, 105
  ethical questions about, 103
  goals of, 75, 76
  moral guidance for, 111-112
Genetic experimentation, nontherapeutic, 94
Genetic manipulation, 102, 145
  challenge of, 105
  ethical and theological aspects of, 101-119

191

obligation of, 114
possible dangers of, 102
social implications of, 145-160
Genetic medicine
 clinical aspects of, **5-51**
 ethical principles and, 87-99
 moral aspects of, **101-112**
Genetic screening; *see* Screening programs
Genetic test, malpractice in performing, 122-124
Genetics, xii
 basic, 4-5
 fundamental, **5-7**
 power of, 101-102
Genome, 105, 107, 108, 112-113
Gildiner v. Thomas Jefferson University Hospital, 126
Gleitman v. Cosgrove, 125
God, 103, 104, 109, 115
 accountability to, 105
 plan of, 110
 union with, 112
Grisez, G., 87
Griswold v. Connecticut, **195**
growth hormome, xi, 82, 107, 147, 151

Handicapped persons, **145-146, 157, 159-161**
Health, promotion of, 94
Health care, cost-effectiveness of, 140-142
Hearing impairment, legal issues on, 125
Heart defects, congenital, 7, 8
 associated with teratogens, 9
Hemoglobin disorders, 10
Hemophilia, **8,** 24, 34
 therapy for, 66

Hereditary deafness, legal issues on, 124-125
Holocaust, 88
Howard v. Lecher, 29, 126, **204-205, 206, 207, 208, 209**
Human, the
 according to developmental approach, **68-71**
 according to genetic approach, **68-71**
 according to social school of thought, **62-64**
 biblical view of, **78-85**
 changes in self view of, **57-59**
 and Christian personalism, **92-97**
 Christian view of, **78-92, 143**
 dignity of, **86-88**
 evolution of, 106
 magisterial teaching and, **85-92**
 meaning of, **55-75, 77-98**
 modes of understanding of, **57-60**
 in moral sense, **60-62**
 philosophical approaches to, **60-74**
 progress and God's plan, **88-89**
 science and, **55-57**
 in scientific sense, **60-62**
 social ramifications of philosophical approaches to, **71-74**
 and suffering, **80-85,** 140
Humanae Vitae, **97, 118-119**
Huntington's disease, **38-42, 124, 129-130, 148**
Hybridization, 82
Hybridomas, 146-147
Hydrocephalus, fetoscopy for, 18
Hydrocephaly, 38
 and pedigree, 34

Hyperphenylalaninemia, 68
Hypothyroidism, 66
  genetic screening for, 50-51
I-cell syndrome, 23-24
Image of God and human, **55-57, 86-88**
Inborn errors of metabolism, 22-24, 64-65, 68-69
Infant mortality, 3
Infertility, 113
Insulin, xi, 107, 147, 151
Intelligence quotient, loss of in phenylketonuria, 67, 68
Interferon, 107, 147, 151
Intuitionism, 89, 94, 96
Inversion, 39

Jacobs v. Theimer, **209**
Jesus Christ, 93, 103
Justice, 87, 136

Kant, E., 91
Karlsons v. Guerinot, **29, 210, 211**
Karyotype, 19
  of father of child with Down syndrome, 21-22
  in mental retardation, 22
Klinefelter's syndrome, **10**

Law, 87
  and genetic disease, **175-231**
Legal aspects of Down syndrome, 124, 126, 129, **206**
Legal issues on genetic diagnostic procedures, 121-130; *see also* Causes of action in "wrongful birth" cases

basis for decisions, 126
California: the *Curlender* case, 121-124, 127, 128-129
New Jersey cases, 125
New York cases, 126
Pennsylvania cases, 126
Turpin v. Sortini, 124-125
Legislative volunteerism, 146
Lesch-Nyhan syndrome, 6-7
  amniocentesis for, 24
Leroy syndrome, 23-24
Liberty, 87
Life, respect of, 93-94

Maher v. Roe, **194, 195**
Malpractice
  in performing genetic test, 122-124
  and standard of care, **201-204**
Maple syrup urine disease, 64-65, 66, 70
Marfan's syndrome, **44-45, 124**
Maritain, 91
Marriage and Church teaching, **115-120**
Marxist analysis of person and society, 91
Material cooperation and sterilization; *see* Sterilization
Maternal age
  and chromosome abnormalities, 11, 19-20, 28
  and Down syndrome, 19-20, **23, 49-50, 55, 124, 148**
Maternal rubella infection, 8
Meckel-Gruber syndrome, 6
Mendel, G., 148-149
Mendelian inheritance, 5
Meningomyelocele
  amniocentesis for, 25, 56

diagnosis of, with
ultrasonography, 17
Mental retardation
associated with citrullinemia,
70
in Down syndrome, 20
and genetic screening, 50
karyotype in, 22
associated with Lesch-Nyhan
syndrome, 7
associated with methylmalonic
acidemia, 24
associated with
Smith-Lemli-Optiz syndrome,
6
associated with teratogens, 9
Mercy killing and
proportionalism, 94
Messenger ribonucleic acid
(mRNA), 78
Methylmalonic acidemia, 24
Microcephaly, 8, 9
Middle Ages, scholastic
philosophy of, 92
Mill, J. S., 90
Miscarriage
associated with amniocentesis,
25-27
early, 15
Molecular biology, 75
Molecular engineering, 147
Mongolism; *see* Down syndrome
Monosomy X; *see* Turner's
syndrome
Moral equivalent, 94
Moral judgment, clinical
experience and, 134-137
Moral values
and counseling, **135-138**
primacy of, **89-92**
Mucolipidoses, 23-24

Multifactorial inheritance, 8
Multifactorial (polygenic)
diseases, 8
Muscular dystrophy, 24, **36-38,
123-124**
mutant gene, 4
Myelomeningocele, 8
Myotonic dystrophy, 33

Nanism, 38
Natural Family Planning (NFP),
21, 126-127, 151, 152
Natural moral law, 91-93, 96
compared with volunteerism,
92
Nazi regime, 88
Negative eugenic programs, 48
Negligence, cases of, **204-208**
Neoeugenics, 148-150
Neural tube defects (NTD), 8, 17,
45-47
amniocentesis for, 25
genetic screening for, 55-57
Neurofibromatosis, **42-44, 124**
legal issues on, 126
Newborn screening, 64; *see also*
Screening programs
for phenylketonuria, 66-67
*N.I.H. Guidelines for Research on
Recombinant DNA
Molecules,* 150
Nitrogen-fixing gene, 156
Nondirective counseling, **26-27,**
142-143
Nondisjunctional chromosome
abnormality, 19-20
Nontherapeutic genetic
experimentation, 94
nucleotides, xi
Nuremberg war trials, 92

Obligation to provide treatment, 142
Obstetrician's duties, **211-214**
Open spine defects, 8
Orchomenos, 58
Organismal engineering, 146
The Origin of the Species by Means of Natural Selection, 148

Parents
  and genetic disease, **120-131**
  moral issues and, **110-111, 120-131**
Park v. Chessin, 126, **181-182, 205, 206, 211**
Paternal and chromosomal abnormalities, 11
Pathogenic organisms, 108
Patient, limitations of, 106
Pedigree, 34-35
Perinatology, 38
Person, primacy of, 116; *see also* Human, the
Personalism, Christian, **92-97**
Personhood, 91
  beginnings of, **69-71**
  and Christian personalism, **92, 97**
  Fletcher's criteria for, **64-65**
    critique of **65-67**
  and magisterial teaching, **85-92**
  meaning of, **62**
  of the unborn child, **69-71**
Phenylalanine, 66, 68
Phenylalanine hydroxylase deficiency, 12
Phenylalanine hydroxylase reaction, 67-68
Phenylketones, 68

Phenylketonuria (PKU), 12, **21, 31-32**
  genetic screening for, 49, 50-51
  treatment for, 66-69
Pierre Robin syndrome, 33
PKU; *see* Phenylketonuria
Planned Parenthood v. Danforth, **196**
Plasmids, 76-78
Polycystic kidneys, 6
  legal issues on, 126
Polygenic inheritance, 8
Pope John XXIII, 115
Pope John Paul II, 95, 111, 116, 117
Pope Pius XII, 105, 106
Postsymptomatic therapy, 63
  for vitamin-responsive disease, 70
Potter syndrome, type 4, 17
Preamniocentesis counseling, 32
Precipitation test, 54
Pregnancy testing, 97
Prenatal diagnosis, 6, 63
  of Down syndrome, 55
  of genetic disease, 80-84
  genetic screening and, 45-60
  of Lesch-Nyhan syndrome, 7
Prenatal diagnostic tools, 16-26
  amniocentesis, 19-26
  amniography, 18
  fetoscopy, 18
  for Meckel-Gruber syndrome, 6
  with ultrasonography, 16-17
Prenatal therapy of genetic disease, 72-73
Presymptomatic therapy, 63
Principle of double effect, **19**
Principle of totality, 106
Procreative decisions, **118-120, 157-158**

195

Procreative power, 94-95
Proportionalism, **93-94**, 95-97
Prospective counseling, 10
Proteins, xi
  amino acid composition of, 107
Providence of God, 106
Prudential personalism, 92, 93

Qualitative test, 46
Quantitative test, 46
Quinlan case, 123

Radiography, **13**
Recombinant DNA technology, 102, 107, 114
  commercialization of, 151-154
  gene splicing, 147
  risk of, 108
Recovery in "wrongful birth cases"
  limited to contraception, **193-195**
  limited to damages from pain and expenses, **187-189**
  limited to special expenses attributable to birth defects, **189-193**
Respect life, 93-94
Responsible stewardship, 105
Retrospective counseling, 10
Ribonucleic acid, messenger, 78
Risks
  for amniocentesis, **16, 18, 20,** 25-26
  for antenatal diagnosis, **18**
  of genetic manipulation, 108-109
Roe v. Wade, **20, 67,** 89, 122, **156, 193, 194, 195, 218, 221**
Rubella infection, maternal, 8

Schizophrenia, 12
Scholastic philosophy, of Middle Ages, 92
Science, 115
  and dominion over nature, 110, 111
  and human, **55-57**
Screening programs, 10, 46-47, 95
  feasibility of, **20-22**
  levels of service in, **22**
  moral questions relating to, **107-110**
  and prenatal diagnosis, 45-60
  problems with, 46
  purpose of, 45
  reasons for, 47-48
  target population in, 46
Scriver, C., 46
Secular humanism, 103
Self-mutilation, 7
Sex selection
  and amniocentesis, **23-24**
  and genetic counselors, **24, 63**
Sherlock v. Stillwater Clinic, **187-188**
Sickle-cell anemia, **7,** 39, 53-55, 84
Single-gene disorders, 4, 5-7
Slavery and natural moral law, 92
Smith-Lemli-Opitz syndrome, 5-6
Smoking, impact of, on pregnancy, 11
Social implications of genetic manipulation, 145-160
Society
  and genetic disease, **155-161**
  responsibilities of, **112,** 133-134, **153-158**
Sonography, **13**
Spastic paraplegia, 33
Species difference, 113

Speck v. Finegold, 126
Speech impairment, legal issues on, 125
Sperm cells, 28
Spina bifida, 141-142; *see also* neural tube defects
  amniocentesis for, 56
  fetoscopy for, 18
  and pedigree, 34
  risk for, 35
Sterilization
  and church teaching, 95, **149, 150**
  and natural moral law, 92
Substitution mentality, **123**
Suffering, 140
  human existence and, **80-85**
  New Testament perspective of, **81-85**
  Old Testament reflections on, **81**
Support systems, 133-134
Supreme Court
  and abortion decisions, 88
  and legislative volunteerism, 91
Surrogate mothers, 146
SV40 tumor virus, 150

Tarasoff v. Board of Regents, **224**
Target population in genetic screening, 46
Tay-Sachs disease 7, 10, **21**, 24, **32-34**, 39, **204-205**
  genetic screening for, 51-53
  legal issues on, 121-124, 126
Technology, 111-112, 115
  development of, 111-112
  and dominion over nature, 110
  ethics for, 116, 117

Ten Commandments, 88-89
Teratogens, 8-9
"Test-tube" babies, 146
Tests for carrier detection, 7
Tetrahydrobiopterin, 68
Thalassemia, 39, 53
Theological aspects of genetic manipulation, 101-119
Thyroid hormone therapy, 66
Totalitarianism, 91
Transculturalism, 92, 96
Translocation, 39
  balanced, **21-22**
  chromosomal defects, 47-49
  4/21, **21**
  21/21 unbalanced, **22**
  unbalanced, 21
Treatment, obligation to provide, 142
Trisomy 13, 6, 19
Trisomy 18, 19, 41, 43
Trisomy 21; *see* Down syndrome
Troppi v. Scarf, **184-186**
Tumor virus, SV40, 150
Turner syndrome, **10**, 34, **61, 71**
Turpin v. Sortini, 124-125
21/21 unbalanced translocation, 22
Type 4 Potter syndrome, 17
Tyrosine, 68

Ultrasonography, 9, 16-17; *see also* Fetal sonography
Unbalanced translocation, 21
United Nations declaration of human rights, 90
United States Constitution, 87
Universe, 109, 111
Utilitarianism, 90

Vaccines, 84
Value(s)
  of genetic counselors, **111-112, 132-138**, 142-143
  human dignity as, **86-88**
  moral, in world of sin, **89**
  of suffering, **80-85**
  progress of humankind as, **88-89**
Vatican Council II, 110, 115
  and common good, **153-155**
  and family, **115-122**
  and human dignity, **86-88**
Virus, SV40 tumor, 150
Vision impairment, legal issues on, 125
Vitamin therapy, 66
Vitamin-responsive disease, 70-73
Vitamins, processing of, 70-71
Voluntarism, 91, 92

Wilson's disease, therapy for, 66
World view
  classical, **57-59**
  modern, **57, 60**
"Wrongful birth" cases, **28, 177, 182-198**
"Wrongful life" cases, **28, 177, 179-182**

X-linked diseases, 24-25
  recessive, **7-8**
X-linked recessive gene, 6
X-ray radiation, 9
XYY males, **10, 43, 61, 71, 228, 229**

Zygote, 4
  nature of, **6**
  and personhood, **69, 71**

RB155 .G386   SIBK
Genetic medicine and engineering : ethic
CARDINAL BERAN LIBRARY
3 3747 00017 8381

DATE DUE			
OCT 10 '89			
FEB 24 '93			

42680

RB
155
.G386
c.1

Genetic medicine and engineering

CARDINAL BERAN LIBRARY

ST. MARY'S SEMINARY

9845 MEMORIAL DRIVE

HOUSTON, TEXAS 77024